The Dilemma of China's Dryland Agriculture in Inner Mongolia

Cilia Neumann

The Dilemma of China's Dryland Agriculture in Inner Mongolia

Economic Growth, Poverty Alleviation and Sustainability –
The Difficulty to Develop the Idea of Environmentalism

Bibliographic Information published by the Deutsche Nationalbibliothek
The Deutsche Nationalbibliothek lists this publication in the Deutsche Nationalbibliografie; detailed bibliographic data is available in the internet at http://dnb.d-nb.de.

Registered: Freiburg (Breisgau), Diss., 2017

Library of Congress Cataloging-in-Publication Data
A CIP catalog record for this book has been applied for at the Library of Congress.

D 25
ISBN 978-3-631-74414-7 (Print) · E-ISBN 978-3-631-74651-6 (E-PDF)
E-ISBN 978-3-631-74652-3 (EPUB) · E-ISBN 978-3-631-74653-0 (MOBI)
DOI 10.3726/b13299

© Peter Lang GmbH
Internationaler Verlag der Wissenschaften
Berlin 2018
All rights reserved.

Peter Lang – Berlin · Bern · Bruxelles · New York ·
Oxford · Warszawa · Wien

All parts of this publication are protected by copyright. Any utilisation outside the strict limits of the copyright law, without the permission of the publisher, is forbidden and liable to prosecution. This applies in particular to reproductions, translations, microfilming, and storage and processing in electronic retrieval systems.

This publication has been peer reviewed.

www.peterlang.com

Acknowledgments

Firstly I want to thank Prof. Dr. Sabine Dabringhaus, my supervisor, for the opportunity to develop this dissertation. I further would like to thank Prof. Dr. Stefan Seitz for his support as a co-supervisor.

I owe my special thanks to Mr. Yun Ting, Dr. David Liu, Prof. Dr. Li Shi and Kaiyun Xie for providing me with all the necessary and important information regarding the potato development in China, especially the potato-projects in Inner Mongolia, Sizwang Qi.

Special thanks to my husband, Dr. Christoph Neumann, for his support, contacts and finally his patience and coaching during the process of developing and writing the following thesis.

I finally would like to thank my friends, family and everyone, who supported me over the period of three years writing the dissertation.

This thesis is dedicated to my children Vico and Golo.

Abstract

Throughout China's history its economy was based on agriculture, mainly on grain production and we are able to retrospect 600 years of rising grain production in China. This was primarily possible by increasing acreage and by raising the yield per acre.[1] From 1700 to 1800, China's population doubled from 160 million to 300 million. At the same time China's production capacity was restricted due to limited arable land. With the explosion of the Chinese population since 1950[2], China's agricultural production could no longer meet the demands. Looking back over the past thirty years China has undergone enormous socio-economic changes of its society and economy. From a socialist-planned economy under Chairman Mao Zedong to the new model of a socialist-market economy introduced under Deng Xiaoping in 1978, China's economic growth has been remarkable. Simultaneously, the encroachment of industries creating high levels of pollution like chemical, coal and rare-earth mining into rural areas has entailed significant economic and environmental costs.[3] The initial damage during the early stages of industrialization in the 1950s until the 1970s caused by heavy industries includes pollution of rivers and lakes as well as air pollution and smog in cities.[4]. But the list of environmental damages since the economic reform era from 1978 forward also contains soil erosion, water shortage, desertification, and salinization-alkalization. Natural disasters like floods and dyke-breaks near to river streams, severe droughts, as well as the phenomenon of the "dzud"[5] together

---

1 Perkins, Dwight H. *Agricultural development in China, 1368–1968*. New Brunswick: AldineTransaction, 2013:13.
2 To illustrate the population increase within 60 years: 1950: 600 Million, in 2013 1.37 billion.
3 Robert F. Ash and Richard L. Edmonds, "China's Land Resources, Environment and Agricultural Production," *The China Quarterly* 156 (1998): 836.
4 Judith Shapiro. *Mao's War against Nature: Politics and the Environment in Revolutionary China*, Studies in Environment and History, Leiden: Cambridge University Press, 2001.
5 Dzud (Mongolian зуд) is the Mongolian term for harsh severe winters, where nearly 10 percent of livestock famishes and dies of starvation due to frozen soils and snow, which make grazing impossible. See: Dzud Natural Disaster, Prevention and Recovery by Elbegdorj Tsakhia (モンゴル国前首相、モンゴル民主党党首). http://www.jst.go.jp/astf/document2/en_33doc.pdf.
Andrew Jacobs, "Winter Leaves Mongolians a Harvest of Carcasses." *New York Times*, May 19, 2010, http://www.nytimes.com/2010/05/20/world/asia/20mongolia.html?_r=0.

complete the picture of Northern China's fragile ecosystems. Since the 1990s until today the increase of environmental distress became a threat to Northern Chinas agricultural productivity.[6] Especially the autonomous province of Inner Mongolia is affected by environmental degradation, which has an adverse effect to Inner Mongolia's economy, which is still based on agriculture, either crop cultivation or livestock and dairy production as one of the economic sectors.[7]

As this thesis is part of a larger project about *Environmental Development in Inner Mongolia*, the research is based on studies in Inner Mongolia. The work is dealing with the more general question, whether economic growth and development is compatible with sustainability and environmental protection in Inner Mongolia. Therefore the possession of environmental awareness is essential to develop pro-environmental behaviour. This thesis specifically examines the development of environmental awareness in terms of protection measures and pro-environmental behaviour among Inner Mongolian farmers and herdsmen. Inner Mongolia is marked by a huge agricultural sector, which represents the fourth industrial pillar of Inner Mongolia's domestic economy, beside energy, metallurgy and food processing.[8] Although Inner Mongolia undergoes the transformation from an economy based on agriculture to an economy based on industry, the majority of the rural population still lives from agricultural production. The ecological situation of Inner Mongolia, severe decline in arable land and alarming grassland degradation aggravates the establishment of a productive and sustainable agriculture. What are the constraints of Inner Mongolia's agricultural problem?

The demand for food will continue to increase due to the growing population while arable land is limited and decreasing. Although China is feeding 20 percent of the world's population, it contains only 10 percent of the world's cultural land.[9] The constant loss in yields will cause already large land requirements to increase the acreage for production, which will in turn result in severe damage to the environment due to land erosion, deforestation and degradation of fertile farmland. Moreover, Inner Mongolia's specific problem lies in its agricultural structure: Dominated by smallholding farms, less technical innovation and an underdeveloped tillage culture with little commitment from the farmers' side for sustainable and modern cultivation methods, Inner Mongolia's agricultural productivity is underperforming. Nevertheless, a sustainable husbandry and

---

6 Ash and Edmonds, "China's Land Resources," 836.
7 China Knowledge Press, *China Business Guide,* Singapore: Chinaknowledge Press, 2004.
8 Ibid.:440
9 *Agricultural Policy Monitoring and Evaluation 2015.* OECD Publishing, 2015.

agriculture are necessary to maintain China's economic growth, while its increasing population simultaneously demands a stable food supply. A domestic staple crop production is essential in terms of China's self-sufficiency and increasing food demand and the drylands of Inner Mongolia are discovered to cultivate the essential staple crops, with the potato leading the way, because of its high nutrition value. The potato has been a major contributor in sustaining the rising food requirements and economic development, as well as helping to alleviate poverty. Therefore, the Chinese government continues to encourage the domestic potato production and is actively supporting the potato's transformation from a dietary choice to a staple food. The government-backed potato scheme strives toward the goal that by 2025 fifty percent of the national potato production will be for domestic consumption.[10]

This dissertation's empirical research in Inner Mongolia concentrates on agriculture and its impact on the environment with particular attention on potato cultivation as part of China's Twelfth Five-Year Plan. By analyzing Inner Mongolia's potato production the approach is to clarify the limitations which hinder a profit-yielding agriculture. Therefore it must be taken into account that the demography of Inner Mongolia is characterized by the Inner Mongolian transition zone, where different economies (*agriculture and animal husbandry*) and people (*Han and Mongols*) collide. Investigating the increase of environmental damage and enhanced economic growth also raised the question of developing environmental awareness in a broader context.

Three hundred surveys were conducted around different counties and regions in Inner Mongolia, questioning farmers as well as herdsmen of different ethnic backgrounds about their understanding of sustainable ecology. Education, culture and motivation are representative for certain factors (demographic, external and internal) and relevant when examining the relation of people to their environment, as these factors influence a person's environmental perception and decision, whether or not they will act pro-environmental. Therefore evaluating the level of environmental perception of the Inner Mongolian people will be a point of relevance in the further discussion about establishing a sustainable agriculture or livestock production in Inner Mongolia.

---

10 "*China to position potato as staple food*," CCTV.com 01–08–2015 04:13 BJT. http://english.cntv.cn/2015/01/08/VIDE1420661517562206.shtml.

Table of Contents

List of Figures XV

List of Maps XVII

List of Tables XIX

List of Abbreviatons XXI

1. **Introduction** 1
   1.1 State of the Field 5
   1.2 Outline and Structure 10
   1.3 Methodology 12
   1.4 Theoretical Framework 14
       1.4.1 Economic Growth and Poverty Alleviation 17
       1.4.2 The Environmental Kuznets Curve 19
       1.4.3 Developing Environmental Awareness 23

2. **Research Framework** 29
   2.1 Project Setting 29
   2.2 Field Research 30
       2.2.1 Challenges 32
       2.2.2 Questionnaire 32
       2.2.3 Realization and Distribution 36

3. **Region of Study** 39
   3.1 Historical Review 41
   3.2 Inner Mongolia 53
       3.2.1 Dryland Insight 57
       3.2.2 Hetao Plain and Hetao Irrigation District 61

4. **Pastoralism versus Agriculture: Different Land-Use Varieties** 71
   4.1 Culture Clash: Han and Mongols – Distinction and Delineation 73
   4.2 Political Changes in Inner Mongolia 79
      4.2.1 Inner Mongolia's Variety of Land-Use 81
      4.2.2 Impact for Mongolian Herders 83

5. **Economy versus Ecology: China's Balancing Act – A Challenge for Inner Mongolia** 89
   5.1 Inner Mongolia as Part of Rural Reforms integrated in China's Agricultural Development 90
   5.2 Economic and Environmental Background of China 96
   5.3 Socio-Economic Situation of Inner Mongolia 106

6. **Empirical Research I: Intensive Cropping in Inner Mongolia: Potato and Poverty Alleviation** 111
   6.1 The Potato's Role in the Twelfth Five-Year Plan, 2011–2015 124
   6.2 The Potato's Role in Inner Mongolia – More than Just a Staple Food 130
   6.3 Case Study 1: Potatoes in Siziwang Qi Banner 133

7. **Inner Mongolia's Environmental Concerns – The Change of the Environment since 1950** 139
   7.1 Irrigation and Water Shortage in Inner Mongolia – A Severe Problem for Farmers and Herdsmen 146
      7.1.1 The Inner Mongolian Water-Rights System and the Function of Water User Associations 154
      7.1.2 Case Study 2: Irrigation in Siziwang Qi Banner 157
   7.2 Desertification and Sandstorms – A Threat to the Local Population 161
   7.3 Dealing with Desertification: National Policies versus NGO Based Action Programs in Inner Mongolia 170

8. Empirical Research II 179

    8.1 Developing Environmental Awareness and Pro-Environmental Behaviour in Inner Mongolia 179

    8.2 Case Study 3 181

        8.2.1 Siziwang Potato Smallholders 183

        8.2.2 Farmers and Herdsmen in Inner Mongolia 190

9. Conclusion 203

10. Bibliography 211

Appendices i

List of Figures

Fig. 1: China's Agricultural Dilemma 4
Fig. 2: Environmental Degradation Kuznets Curve 20
Fig. 3: Environmental Degradation Tunnel Effect Kuznets Curve 22
Fig. 4: Development of China's Economy 98
Fig. 5: Potato Producing Countries in the World 2012 118
Fig. 6: Environmental Awareness and Action Building Process in
       Inner Mongolia 180
Fig. 7: China's Education System 187
Fig. 8: Crop Distribution of Interviewees in Inner Mongolia 191

List of Maps

Map 1: Prefectures of Inner Mongolia 40
Map 2: Russia, China and the Origins of the People's Republic of Mongolia 50
Map 3: Larger Area of Hetao (including Houtao, Qiantao and Xitao)
during Ming and Qing Dynasty 62
Map 4: Agricultural Areas in Hetao during Ming and Qing Dynasty 63
Map 5: Hetao Irrigation District with its Canal System 65
Map 6: China's, Agro-Ecological Zones of Potato Cultivation 131
Map 7: Precipitation Distribution China 147
Map 8: Expansion of Northern Chinese Drylands 1948–2008 167
Map 9: Vegetation Cover in North China 168

List of Tables

Tab. 1:	Fishbein & Ajzen 1975	24
Tab. 2:	Rajecki 1982	25
Tab. 3:	Hines, Hungerford & Tomera 1986	25
Tab. 4:	Fietkau & Kessel 1981	27
Tab. 5:	Blake 1999	28
Tab. 6:	Structure of Questionnaire	33
Tab. 7:	Coding and Categorizing	35
Tab. 8:	Composition of Inner Mongolia's GDP	107
Tab. 9:	Potato Planting Area, Production and Yield 2006	122
Tab. 10:	Potato Production in China 1961–2009	123
Tab. 11:	Crop Rotation and Intercropping among Thirty Farmers in Siziwang Qi	134
Tab. 12:	Crop Selection among Thirty Farmers in Siziwang Qi	135
Tab. 13:	Purpose of Re-Investment among Thirty Farmers in Siziwang Qi	136
Tab. 14:	Field Size among Thirty Farmers in Siziwang Qi	136
Tab. 15:	Principles of Water User Associations	156
Tab. 16:	Crop Distribution among Thirty Farmers in Siziwang Qi	157
Tab. 17:	Soil Issues among Thirty Farmers in Siziwang Qi	159
Tab. 18:	Allocation of Re-Investment among Thirty Farmers in Siziwang Qi	160
Tab. 19:	Level of Graduation	193
Tab. 20:	Correlation of Education to Awareness and Pro-Environmental Behaviour	195
Tab. 21:	Effect of Environmental Awareness to Future Perspective	199
Tab. 22:	Parents Career Choices for their Children	200

List of Abbreviatons

AD	*Anno Domini* [Latin]; after Christ
AHNI	Annual Household Net Income
BC/BCE	Before Christ/Before the Common Era
CCICCD	China Coordination Implementation Committee to Combat Desertification
CCICED	China Council for International Cooperation on Environment and Development
CEO	Chief Executive Officer
CIP	Centro Internacional di Papa (International Potato Center)
COD	Chemical Oxygen Demand
EKC	Environmental Kuznets Curve
FAO	Food and Agriculture Organization of the United States
GDP	Gross Domestic Product
GfG	Grain for Green Policy
HID	Hetao Irrigation District
HPRS	Household Production Responsibility System
HRS	Household Responsibility System
ID	Irrigation District
NGO	Non-Governmental Organization
NPO	Non-Profit Organization
OECD	Organization for Economic Co-operation and Development
OISCA	Organization for Industrial, Spiritual and Cultural Advancement-International
PRC	People's Republic of China
PSE	Producer Support Estimate
R&D	Research and Development
RMB	Renminbi, Currency China
ROI	Return of Investment
SFAGM	Small Farmers Adapting to Global Markets
UN	United Nations
UNCCD	United Nations Convention to Combat Desertification
US/USA	United States/United States of America
USDA	United States Department of Agriculture

WHO	World Health Organization
WTO	World Trade Organization
WUA	Water User Association
YRWCC	Yellow River Water Conservancy Commission, China

"The straight-trunked tree is the first to be felled; the well of sweet water is the first to run dry"
    Zhuangzi, Third Century BCE

1. Introduction

To cope with the needs of its growing population, mainly to guarantee food security, China must increase its agricultural production and yields of staple crops. This leads to ever-increasing agricultural land-use in Inner Mongolia, and in the worst cases, remaining grassland or other eco-systems are being transformed into farmland. Inner Mongolia already suffers from land fragmentation due to small-farm households and the cultivation of "towel-size" fields.[11] The government forcefully tries to limit these land-use practices with land consolidation in order to increase the agricultural productivity, household income and crop yield as well as to improve land quality. For the agricultural and pastoral sector the chronic water shortage of Inner Mongolia, which have led to eroded fields and the expansion of desertificated acreage, is a major problem hindering productivity. Therefore methods to improve agricultural water-use efficiency are of top priority for the government as well as for the agricultural stakeholder.

Observing the situation of Inner Mongolia's ecology, a specific sequence of events have contributed to the environmental problems visible today. These include the development of intensive crop cultivation as well as a modernized livestock production, the vast economic development and the involvement in globalized markets which have all left a measurable mark on Inner Mongolia's environment. Looking at the current situation of China's agriculture, the *Twelfth Five-Year Plan* appears to be very ambitious, and it is questionable how the government and farmers will be able to implement and support the new policy.

Governmental policies – still with emphasis on economic growth – rely on the model of poverty alleviation via rising productivity, acreage and yield. The prevention of soil erosion, acidification and desertification are of national priority and supported land consolidation appears to be a potential tool to reduce further field fragmentation.

In summary the author argues that Inner Mongolia's dryland agriculture is in a dilemma. It is confronted with environmental problems such as severe soil erosion, salinization and desertification, all due to chronic water shortages and agricultural mismanagement, which hampers agronomical development.

---

11 The average size of a potato field in China is 2.2 hectares. Smallholders, cultivating less than 1mu (1ha is 15mu) are a main cause for the above mentioned land fragmentation.

Three aspects are of particular interest and are important for a sustainable, profitable economic development in Inner Mongolia and therefore will now be addressed.

1. *How do the past and current food resources meet the demand of China's population?*

Potato production is part of China's national policy to become self-sufficient in the major food crops, despite its limited agricultural resources.[12] Therefore the productivity and yield has to increase to meet the nation's demand of a stable food supply. But the realities show that Inner Mongolia's potato production is not successful in terms of profitability. On the contrary, the production is marked by substandard yields and increasing acreage.

➔ What are the key factors responsible for the crop loss in Inner Mongolia? Western potato growers are extraordinary efficient and harvest more than 40 metric tons per hectare, whereas an Inner Mongolian farmer has a yield of 15 metric tons per hectare.[13]

➔ How does Inner Mongolia's cultivation method affect the environment when arable land is limited but the production has to increase?

Because Inner Mongolia produces more than forty percent of China's total domestic potato production its cultivation is of high priority for China's government in this region. The potato is particularly significant because of its quality in terms of yield, drought-resistance, and high nutritional value. By supporting and increasing potato production and potato processing industry in Inner Mongolia, China has the opportunity to establish a sustainable, but still highly productive agricultural system, able to compete with international standards and the world market. But the majority of the rural population in Inner Mongolia is still living under poor conditions and their main concern is to improve living standards, achieve higher income and gain access to the local markets to sell their products. Thus, reflections upon environment or sustainability are of a lower priority.

---

12 OECD. "OECD-FAO Agricultural Outlook 2013–2022: Highlights." 2013. Accessed February 15, 2015.
13 "Our Crop Focus: Looking at the R&D Innovations behind some of our solutions in Diverse Field Crops and Speciality Crops." Science matters Unpublished manuscript, last modified January 26, 2016, p. 13–14. http://www.nxtbook.com/syngenta/Syngenta_Research_and_Development/Autumn_2013/index.php?startid=13#/0. Keeping up to date with Syngenta Research and Development.

Furthermore, the possibilities for these farmers to invest in good quality seed, fertilizer and crop protection or chemical treatments are limited. These difficulties may altogether result in the destruction of natural resources as well as overuse and infertile soils, especially through the improper use of pesticides and fertilizers.

2. Economic pressure causes environmental problems – How can Inner Mongolia address this problem?

Inner Mongolia's agricultural productivity is in stagnation due to a lack of sufficient available arable land and an increasing population and demand for products. The factor preventing the increase of productivity is the current situation facing Inner Mongolian farmers. Mainly the smallholders cultivating the land, without access to technological innovations, which are practicing an underdeveloped tillage culture and show little commitment to the business of farming. This has a direct impact on the environment and causes a domino-effect. Thus, environmental degradation will increase as long as agriculture is resource extracting and not sustainable in terms of profitability. The ecological situation of China, specifically an increase of arid and semi-arid areas which are less suitable for intensive agriculture, leads to additional pressure on the limited arable land. As a consequence, the problem of exhausted soils and degradation of cultivated land will rise as well. Therefore China must modernize its agricultural production in order to guarantee productivity growth and sustainable management of its fragile ecology.

➜ How do the demographics of Inner Mongolia's influence this development?

Looking back on a history characterized by in-migration processes the Inner Mongolian agriculture arose out of Han-Chinese tillers, who started the cultivation of the fertile pastures. The Han-Chinese influence has created a melting pot of Han-Mongolian co-existence in both agriculture and animal husbandry. Today neither the nomadic lifestyle (Mongolian), nor farming household (Han-Chinese) are distinguishable from its ethnic background.

Fig. 1: China's Agricultural Dilemma

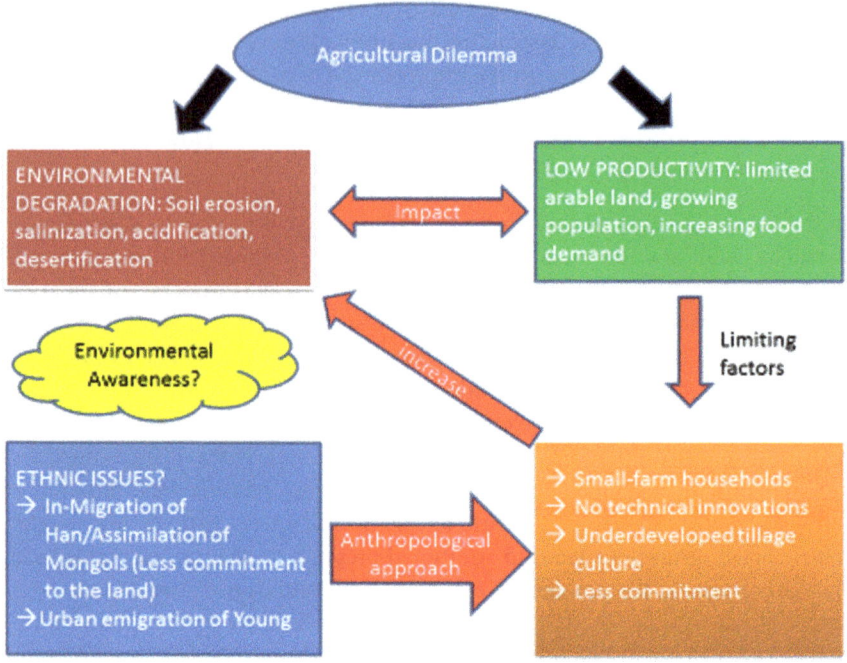

Source: Cilia Neumann (2016), *China's Agricultural Dilemma*

The approach of this dissertation is to provide insight into Inner Mongolia's dryland agriculture, its agitators, environmental, social and political obstacles as well as opportunities.

3. Developing environmental awareness – How can Inner Mongolians contribute to pro-environmental behaviour?

To develop sustainability, either in agriculture or in husbandry, the factor of attachment to the homeland seems to be of interest. Sustainable behaviour appears to only be effective when the habitat is respected. Often hunters and gatherers serve as an example for societies which manipulate their environment to a relatively little extent.[14] To clarify this specific theory, the presumption is that all

---

14 Sutton, Mark Q.; Anderson, E. N. 2014: *"Introduction to cultural ecology"*. 3. ed. Lanham, Md.:AltaMira.

humans are aware of changing and manipulating their environment to achieve a desired goal. But obviously, someone with a bulldozer influences the environment to a greater severity than someone with a stone axe.[15] That awareness of manipulating interactions with the environment is based on the assumption that some societies perceive their environment as something they live from, because they do maintain a relation with their natural habitat, which implies pro-environmental behaviour.[16]

But what happens when people are forced to migrate and settle on new territory with little sentiment for their land or with no relation? Hence, questioning Inner Mongolians about their ecological awareness can provide answers as to whether they feel responsible for their actions or not. Further the question might answer, if the agricultural dilemma in Inner Mongolia is a local problem, connected to economic issues or whether it is a global phenomenon which most developing countries face where environmental awareness only develops with rising living standards.[17]

Nevertheless, Inner Mongolian agriculture is no longer an independent economic sector. On the contrary, it serves national needs, is embedded in global markets and national policies try to regulate the economy, as well as prescribing environmental guidelines for its development in Inner Mongolia.

1.1 State of the Field

Research on Chinese environmental history is relatively new and not widespread. Academics dealing with environment and ecology, who concentrate on China or South-East Asia, are typically focused on certain aspects like the increase of water-pollution in rivers and irrigation basins.[18] Alternatively, they deal

---

15 Ibid.:117
16 Ingold, Tim. *The perception of the environment: Essays on livelihood, dwelling and skill.* Repr. London: Routledge, 2008, 43.
17 Huang, Pingsha, Xiuli Zhang, and Xingdi Deng. "Survey and analysis of public environmental awareness and performance in Ningbo, China: a case study on household electrical and electronic equipment." *Journal of Cleaner Production* 14, no. 18 (2006): 1635–43.
18 Stephenson, David, and Margaret S. Petersen. *Water resources development in developing countries* 41. Amsterdam, New York, New York, NY, U.S.A: Elsevier, 1991.
   Michael P. Lawrence. "Damming Rivers, Damning Cultures." *American Indian Law Review* Vol. 30, no. 1 (2005/2006): 247–89. http://www.jstor.org/stable/20070754.
   Xu, Xu, Guanhua Huang, Zhongyi Qu, and Luis S. Pereira. "Assessing the groundwater dynamics and impacts of water saving in the Hetao Irrigation District, Yellow

with air pollution, CO^2 emissions and the causes and effects of smog in China's megalopolises.[19]

Much of the research related to this topic and conducted in the past fifteen years concentrates largely on the Yellow River and the phenomenon of rising silt levels and its historical potential for severe flooding.[20] In particular, the works of Jiongxin Xu (Institute of Geographical Sciences and Natural Resources Research, Key Laboratory for Water Cycle and Related Land Surface Processes, Chinese Academy of Sciences, Beijing, China) should be mentioned.[21] Additionally, much research on irrigation issues and environmental damage has been published in

---

River basin." *Agricultural Water Management* 98, no. 2 (2010): 301–13. Accessed June 12, 2013. doi:10.1016/j.agwat.2010.08.025. http://www.elsevier.com/locate/agwat.

19 Cheng, Zhiming, Mark Wang, and Junhua Chen, eds. *Urban China in the New Era: Market Reforms, Current State, and the Road Forward.* Berlin, Heidelberg, s.l.: Springer Berlin Heidelberg, 2014. http://dx.doi.org/10.1007/978-3-642-54227-5.

Koppel, Bruce, Norton S. Ginsburg, and T. G. McGee, eds. *The Extended metropolis: Settlement transition in Asia.* Honolulu: University of Hawaii Press, 1991.

Young, Jason. *China's Hukou System: Markets, Migrants and Institutional Change.* New York: Palgrave Macmillan, 2013. http://gbv.eblib.com/patron/FullRecord.aspx?p=1249600.

20 Liu, Guo W. "On the geo-basis of river regulation in the lower reaches of the Yellow River." *Science China, Earth Sciences* Vol. 55, no. 4 (2012): 530–44.

Liu Changming and Zhang Shifang. "Drying Up of the Yellow River: Its Impacts and Counter-measures: Mitigation and Adaptation Strategies for Global Change 7: 203–214, 2002." *Mitigation and Adaptation Strategies for Global Change* Vol. 7 (2002): 203–14.

Mark Giordano, Zhongping Zhu, and Ximing Cai. "Water Management in the Yellow River Basin: Background, Current Critical Issues and Future Research Needs." Accessed June 11, 2013.

Wang, Guangqian, Baosheng Wu, and Zhao-Yin Wang. "Sedimentation problems and management strategies of Sanmenxia Reservoir, Yellow River, China." *Water Resources Research* 41, no. 9 (2005): n/a.

Wu, Baosheng, Zhaoyin Wang, and Changzhi Li. "Yellow River Basin Management and current issues." *Journal of Geographical Sciences* Vol. 14 (2004): 29–34. http://www.geog.cn.

21 Xu, Jiongxin. "Growth of the Yellow River Delta over the past 800 Years, as Influenced by Human Activities." *Geografiska Annaler. Series A, Physical Geography* Vol. 85, no. 1 (2003): 21–30.

Xu, Jiongxin. "Historical bank-breaching of the lower Yellow River as influenced by drainage basin factors." *Catena* Vol. 45 (2001): 1–17. http://www.elsevier.com/locate/catena.

the course of the construction of the Sanmenxia Dam (三峡大坝).[22] Both the construction of the dam and the discussion and research about it gained worldwide publicity.[23]

Further fields dealing with environmental alteration in China are focusing on the development of grassland degradation in the past fifty years. In particular, research has been undertaken in Inner Mongolia and the independent state of Mongolia[24], as the two regions are famous for their huge grasslands as well as for their problem of fighting against severe grassland deterioration. Many scholars have examined the different stages of grassland destruction caused by

---

Xu, Jiongxin. "Naturally and Anthropogenically accelerated sedimentation in the Lower Yellow River, China, over the Past 13.000 years." *Geografiska Annaler. Series A, Physical Geography* Vol. 80 A, no. 1 (1998): 67–78.

Xu, Jiongxin. "A Study of Long Term Environmental Effects of River Regulation on the Yellow River of China in Historical Perspective." *Geografiska Annaler. Series A, Physical Geography* Vol. 75, no. 3 (1993): 61–72. http://www.jstor.org/stable/521025.

22 Wang, Guangqian, Baosheng Wu, and Zhao-Yin Wang. "Sedimentation problems and management strategies of Sanmenxia Reservoir, Yellow River, China." *Water Resources Research* 41, no. 9 (2005): n/a.

Baosheng Wu, M.ASCE, Guangqian Wang, and and Junqiang Xia. "Case Study: Delayed Sedimentation Response to Inflow and Operations at Sanmenxia Dam." *JOURNAL OF HYDRAULIC ENGINEERING*, 2007, 482–94.

23 The author refers to the works of Dai Qing (戴晴), a Chinese activist and journalist, who draws attention to issues, related to the Three-Gorges-Dam. The book "Yangtze, Yangtze!" was banned after the Tiananmen Square Protests (天安門屠殺 June 4 1989), after it was first published in 1989. At the same time, Dai Qing was denounced and imprisoned for sharing her critical opinions against the Three-Gorges-Dam project.

Dai, Qing, Patricia Adams, and John Thibodeau. *Yangtze! Yangtze!* English ed. London, Toronto: Earthscan, 1994.

Qing, Dai, John Thibodeau, and Philip B. Williams. *The river dragon has come! The Three Gorges dam and the fate of China's Yangtze River and its people.* An East Gate book. Armonk, N.Y.:M.E. Sharpe, ©1998.

24 Jeremy Swift, and Robin Mearns. "Mongolian pastoralism on the treshold of the twenty-first century." *Nomadic Peoples*, no. 33 (1993): 3–7. Accessed July 1, 2013.

Kakinuma, Kaoru, Takahiro Ozaki, Seiki Takatsuki, and Jonjin Chuluun. "How Pastoralists in Mongolia Perceive Vegetation Changes Caused by Grazing." *Nomadic Peoples* 12, no. 2 (2008): 67–73.

Dennis P. Sheehy. "Grazing management strategies as a factor influencing ecological stability of Mongolian grasslands." *Nomadic Peoples*, no. 33 (1993): 17–30. Accessed July 1, 2013.

Marin, Andrei. "Between Cash Cows and Golden Calves: Adaptations of Mongolian Pastoralism in the 'Age of the Market'." *Nomadic Peoples* 12, no. 2 (2008): 75–101.

overgrazing, soil erosion, and desertification as well as acreage expansion into the grassland regions.[25] Often the correlation between climate induced environmental changes and degradation of former fertile grasslands is mentioned.[26] The Chinese government also maintains several institutes and research departments with international cooperation concerning the development of severe grassland deterioration in Inner Mongolia.[27] Specialized research methods, such as using a remote, satellite sensing system for monitoring grassland deterioration as well as desertification and soil erosion in agricultural areas of Inner Mongolia, have been part of geographical, geological and archeological studies.[28] There are additionally environmental research programs dealing with the topic of sustainability

---

    Xiuxia, Bao, Yi Jin, Liu Shurun, Gaowa Jimuse, Wang Puchang, and Lian Yong. "Effects of Different Grazing on the Typical Steppe Vegetation Characteristics on the Mongolian Plateau." *Nomadic Peoples* 12, no. 2 (2008): 53–66.

25 Zhao, Ying, Stephan Peth, Rainer Horn, Julia Krümmelbein, Bettina Ketzer, Yingzhi Gao, Jose Doerner, Christian Bernhofer, and Xinhua Peng. "Modeling grazing effects on coupled water and heat fluxes in Inner Mongolia grassland." *Soil and Tillage Research* 109, no. 2 (2010): 75–86. Accessed June 11, 2013.

    Jun Li, Wen, Saleem H. Ali, and Qian Zhang. "Property rights and grassland degradation: A study of the Xilingol Pasture, Inner Mongolia, China." *Journal of Environmental Management* 85, no. 2 (2007): 461–70.

    Ye, Yu, and Xiuqi Fang. "Boundary shift of potential suitable agricultural area in farming-grazing transitional zone in Northeastern China under background of climate change during 20th century." *Chinese Geographical Science* 23, no. 6 (2013): 655–65.

26 Barthold, F.K., M. Wiesmeier, L. Breuer, H.-G. Frede, J. Wu, and F.B. Blank. "Land use and climate control the spatial distribution of soil types in the grasslands of Inner Mongolia." *Journal of Arid Environments* 88 (2013): 194–205. Accessed June 11, 2013.

    Qian, S., L. Y. Wang, and X. F. Gong. "Climate change and its effects on grassland productivity and carrying capacity of livestock in the main grasslands of China." *The Rangeland Journal* 34, no. 4 (2012): 341.

27 Staff, National R. C. *Grasslands and Grassland Sciences in Northern China // Grasslands and grassland sciences in northern China: A report of the Committee on Scholarly Communication with the People's Republic of China, Office of International Affairs, National Research Council.* Washington: National Academies Press; National Academy Press, 1992.

28 Hu, Z. M., S. G. Li, J. W. Dong, and J. W. Fan. "Assessment of changes in the state of the rangelands of Inner Mongolia, China between 1998 and 2007 using remotely sensed data." *The Rangeland Journal* 34, no. 1 (2012): 103.

    Yu, Ruihong, Tingxi Liu, Youpeng Xu, Chao Zhu, Qing Zhang, Zhongyi Qu, Xiaomin Liu, and Changyou Li. "Analysis of salinization dynamics by remote sensing in Hetao Irrigation District of North China." *Agricultural Water Management* 97, no. 12 (2010): 1952–60. Accessed June 11, 2013.

and the protection of the last unimpaired grasslands and ecosystems of Inner Mongolia.[29]

Rural Development focusing on nomadic and pastoral societies has been discussed since the 1990s. The works of John W. Longworth (Emeritus Professor of the University of Queensland, Australia), who specialized in economics and agriculture of China, with emphasis on animal husbandry, have been essential to this dissertation.[30] The impact intensive agriculture has on the local environment is discussed by several scientific institutes and departments, the *Organization for Economic Co-operation and Development* (OECD) and certain agribusinesses, which support programs examining the agricultural sector, and are based on the concept of Research and Development.[31] Research studies dealing with the cultivation of potatoes in China (Inner Mongolia) are still supported by the CIP (*Centro Internacional di Papa*) in Beijing. The program is strongly promoted by the Chinese government, which further announced (January 2014) the establishment of the potato as a staple food in China.[32]

Lee, R., F. Yu, K. P. Price, J. Ellis, and P. Shi. "Evaluating vegetation phenological patterns in Inner Mongolia using NDVI time-series analysis." *International Journal of Remote Sensing* 23, no. 12 (2002): 2505–12. Accessed June 11, 2013.

29 Ellis, J. E. et al. "Sustainability of Inner Mongolian Grasslands: Application of the Savanna Model." *Journal of Range Management* 56, no. 4 (2003): 319–28. Accessed March 30, 2014.
D. R. Kemp and D. L. Michalk, ed. *Development of sustainable livestock systems on grasslands in north-western China: Proceedings of a workshop held at the combined International Grassland Congress and International Rangeland Conference, Hohhot, Inner Mongolia Autonomous Region, China*. 2008. Accessed March 19, 2014.

30 Longworth, John W., ed. *China's rural development miracle: With international comparisons papers presented at an international symposium held at Beijing, China, 25–29 October 1987*. St. Lucia, Qld, Portland, Or: University of Queensland Press; Distributed by International Specialized Bk. Services, 1989.
John W. Longworth, John W. Longworth, and Gregory J. Williamson. *China's pastoral region: Sheep and wool, minority nationalities, rangeland degradation and sustainable development*. Wallingford: CAB Internat. [u.a.], 1993.

31 "Environment, water resources and agricultural policies: Lessons from China and OECD countries; [these proceedings bring together papers from the Workshop on Environment, Resources and Agricultural Policies in China, held in Beijing, on 19–21 June 2006]." Paris.

32 http://www.potatogrower.com/2015/01/china-pushing-potatoes.

Personal Approach

This thesis with focus on the development of environmental awareness and pro-environmental decision-making and behaviour is embedded in the broader discussion about environmentalism. Therefore the research conducted in Inner Mongolia provides insight in the development process of a sustainable agriculture and the evolution of environmental awareness in an economically and ecologically underdeveloped region.

The strength of this research lies in the fact that every human perceives its environment and maintains an individual relation with it. Thus questions on environmental changes produces results in terms of environmental attitudes and values as well as environmental knowledge and education, which might draw conclusions about general awareness on environmental issues.

However, this research also involves weaknesses regarding objectivity/subjectivity on the topic. As the informant talks about his or her perceptions on environmental change or degradation, the answers provided are not objective. They have to be correlated to the actual environmental problems occurring within the habitat, the person lives in.

Additionally the implementation of pro-environmental behaviour depends on several internal and external factors.[33] Without knowing about these aspects, the answers provided in the interviews are not useful. Therefore information on infrastructure, political and cultural factors and economy are part of the questionnaire and in combination visualize the environmental concern of the Inner Mongolian people.

1.2 Outline and Structure

This work is divided into eight parts. The introduction provides a brief overview of China's economic development over the last decades and encapsulates the economic and environmental background of Inner Mongolia. An introduction into methodology will be given and the theoretical framework for analyzing the correlation of economic growth, poverty alleviation and environmental degradation will be illustrated and explained with reference to the model of *the Environmental Kuznets Curve* by Grossman and Krügers (1991). The question of

---

33 Kollmuss, Anja, and Julian Agyeman. "Mind the Gap: Why do people act environmentally and what are the barriers to pro-environmental behavior?" *Environmental Education Research* 8, no. 3 (2002): 239–60. Accessed April 22, 2015. doi:10.1080/13504620220145401.

developing environmental awareness is explained by different models (*Fishbein & Ajzen 1975, Fietkau & Kessel 1981, Rajecki 1982, Hines, Hungerford & Tomera 1986 and Blake 1999*) which help to demonstrate the correlation of education and awareness, which leads to pro-environmental behaviour.

The second chapter outlines the research framework. A general overview of the study area with a short discussion of relevant literature provides a summary on the geography and ecology of Inner Mongolia. Challenges and difficulties regarding the questionnaire, its design and distribution of the empirical research will additionally be outlined and described in chapter two.

The third chapter provides detailed insight into the region of the study. The historical overview will illustrate the development of Inner Mongolia as an economic and social transition zone. Geographical particularities like drylands and irrigated districts point out that Inner Mongolia has diverse environments, which have to be managed individually. Emphasis has been placed on the features of Inner Mongolia's grassland and dryland agriculture and the irrigated Hetao area.

Chapter four and five deal with the socio-economic situation of Inner Mongolia. Focus has been placed on the remarkable circumstance of Inner Mongolia as a transition zone of ethnic and economic interaction where different groups of people and traditional economies collide. As main actors of the transformation process of Inner Mongolia's environment, insights into their particular histories, cultures and economies will be examined. Because of this combination of Han and Mongol influences, the clash of agriculture and animal husbandry is ubiquitous in Inner Mongolia. The sociological impact of a multi-ethnic society and the influence of Inner Mongolia's political history on the environment will also be analyzed. The early settlement and social policies, which influenced further development of Inner Mongolia, cannot be considered separately.

The empirical research is divided into three sections and embedded in chapter six, seven and eight. Chapter six, including the first empirical narratives, focuses on agriculture and its impact on the environment with thematic priority being placed on potato cultivation in Inner Mongolia as a part of Chinas Twelfth Five-Year Plan, which had the objective to promote Inner Mongolia's rural development. Inner Mongolian potato farmers face economic and ecological challenges, which threaten their productivity and crop yield. Strategies, political guidelines and global economic alliances, which influence the Inner Mongolian potato production are represented and discussed.

Chapter seven investigates the increase of environmental damage since the 1950s. Determinants for environmental degradation are assessed in the context of Inner Mongolia's dryland agriculture. With the help of secondary literature, detailed attention has been given to the ecological alteration and its related phenomenon of water shortage and desertification in Inner Mongolia. To deal with the water-scarcity, Water User Associations (WUA) as a self-administration tool play an important role in terms of water-resource regulation in Inner Mongolia. Space is left for the farmers in Siziwang to explain their specific problems related to irrigation and how they deal with it. Analyzing the expansion of drylands, specific focus is given to the interaction of government, NGOs and local population, especially with due regard to different afforestation projects in Inner Mongolia.

Finally, chapter eight unveils the difficulties of developing environmental awareness in a broader context. Different models (*Fishbein & Ajzen 1975, Fietkau & Kessel 1981, Rajecki 1982, Hines, Hungerford & Tomera 1986, Blake 1999*) have attempted to explain the gap between the correlation of environmental knowledge and awareness in Inner Mongolia and demonstrating pro-environmental behaviour of farmers and herdsmen. Both the attitude towards environment and the individuals' behaviours are key factors for the development of environmental awareness. The assumption is that the existence of environmental awareness is responsible for less commitment or higher engagement in pro-environmental behaviour. The theoretical approach of the different models may help to illustrate the complexity of Inner Mongolia's relatively slow development in terms of environmentalism. This chapter also summarizes the results of the case studies and contextualizes this information within the history and development of environmental awareness. Consequently, the strengths and weaknesses of the research addressed in the introduction will be revisited. The requirement for establishing a sustainable ecological balance in China, respective of Inner Mongolia and in association with a growing flourishing economy, lies in an active, reflective civil-society, accepting the liability for their actions. Raising awareness of ecology and environmental damage is a starting point.

1.3 Methodology

Regarding the methodical approach of this thesis, the author was confronted with certain problems. Specifically the problem of working with sufficient documentary material in Chinese was related to the fact that the access was limited, either due to the sensitivity of the topic or that they were predominantly written in English. Especially most of the primary sources connected with environmental deterioration, concentrating on desertification of the grassland steppe ecosystem or the problems of controlling the Yellow River were published in

English.³⁴ The assumption is that the material was released in conjunction with international conferences dealing with the issue or for international audience and therefore mainly accessed via English written journals. However the authors are Chinese scientists, working in China in national institutes or organisations and hence the author had to rely on their provided information.

In general the sources for this project were gathered and researched primarily in Inner Mongolia, with some literature work being conducted in Beijing. Primary sources from the National Archives in Beijing and Provincial Archive in Hohhot as well as from University libraries in Inner Mongolia and Beijing were invaluable to this work, particularly in chapters three, four and five. Secondary sources for this work included newspapers and local chronicles. Newspapers before 1987, concentrating on the period of two major political campaigns (1958–1960 *Great Leap Forward* and 1966–1976 *Great Proletarian Cultural Revolution*) were especially interesting in their reporting related to the Inner Mongolian environment. Yearbooks and local chronicles based on regional history; geography and economic development (Baotou/Bayan Nur/Wulateqi) have been evaluated by searching for chapters dealing with environment, degradation, and agricultural development and grassland issues.

Certain information dealing with the development of intensive potato production in Inner Mongolia was provided by a Swiss agro-business company. The informants granted me access to business-relevant information regarding the development of the potato-market in Inner Mongolia. Due to conflict of interest, as the informants are job and company committed, the information provided had to be handled confidentially. Open access concerning this specific business and development was provided by the International Potato Center (*Centro Internacional di Papa*) in Beijing, and the Agricultural Bureau in Hohhot.

---

34 Wei, Liu, Cao Shengkui, Xi Haiyang, and Feng Qi. "Land use history and status of land desertification in the Heihe River basin." *Natural Hazards* 53, no. 2 (2010): 273–90.
Wu, Bo, and Long J. Ci. "Landscape change and desertification development in the Mu Us Sandland, Northern China." *Journal of Arid Environments* 50, no. 3 (2002): 429–44. Accessed June 11, 2013.
Kang, L., X. Han, Z. Zhang, and O. J. Sun. "Grassland ecosystems in China: review of current knowledge and research advancement." *Philosophical Transactions of the Royal Society B: Biological Sciences* 362, no. 1482 (2007): 997–1008. Accessed June 11, 2013. doi:10.1098/rstb.2007.2029.
Zhang, Dian, and Changxing Shi. "Sedimentary Causes and Management of Two Principal Environmental Problems in the Lower Yellow River." *Environmental Management* 28, no. 6 (2001): 749–60.

Furthermore, informal interviews were conducted and many conversations with industry insiders were shared.

Beyond the previously mentioned sources, two variations of structured questionnaires were created for the evaluation of local attitudes towards their environment. These were prepared in advance in both English and Chinese. The standardized interviews included the same framework of questions so that the answers could easily be compared with each other. One variation of the questionnaire was undertaken with local farmers and herdsmen of different ethnic background (Han, Mongol, and Hui) and distributed through students from the Inner Mongolian Agricultural University in Hohhot. Two hundred and sixty-eight interview sheets were distributed in different counties and regions in Inner Mongolia and thirty exclusively in Siziwang Qi Banner focusing on the potato farming sector. All questionnaires included the principle of open questions. The narratives included sections for general data (age, sex, family status, children and education…) and specific questions regarding the family's business as well as the most interesting, but also problematic and sensitive part including open questions on environmentalism. The narratives were in Chinese, to guarantee that the participants can answer in their mother tongue.[35] During the preliminary stages the questionnaires were constructed by myself and afterwards double-checked by the supervisor. Additionally the author discussed the sensitive parts and issues with native Chinese colleagues, to guarantee that the informants are able to understand the content matter. All answers were handwritten in Chinese and afterwards translated into English and archived in Excel sheets to simplify later evaluation. In addition, three non-standardized interviews were completed. The interviews were held and manually recorded in Chinese. Assistance during the interviews was provided by a Chinese native, who supported the translation and helped to clarify any issues regarding language barriers. These materials were later translated into English and archived in a fieldwork-diary.

1.4 Theoretical Framework

Regarding the theoretical framework of this thesis a brief outline will be given. In this section, certain terminology which will be employed throughout this work will be clarified. Specifically, the terms *environment* and *nature* will be explained

---

35 The primary assumption that the participants may not speak and write Chinese has not been confirmed, as the people in Inner Mongolia, regardless of their ethnic descent master the Chinese language. Aside from that, Chinese was the only language advantageous for both parties, the interviewer and the participant, to communicate with each other.

and how this terminology has been understood. In the following three theories will be discussed, independently of each other dealing with environmental deterioration. Finally, the development of environmental awareness and how this seems to be connected with education will be detailed explained and analyzed.

The academic background for this thesis is based on Anthropology and Sinology. The empirical research also followed the *Environmental Economics* approach[36], which examines the impact of governmental policies and the grade of environmental destruction in relation to behaviour and awareness of the local population of Inner Mongolia.

The term *environment* must be clarified and disassociated from the term *nature*. Nature is the phenomena of the physical world, including all species, landscapes and other features and products of the earth, as opposed to everything which is manmade[37], whereas environment refers to the condition and habitat a species lives in.[38] Environmental damage or deterioration in general refers to the human impact and the relation between humans and their environment.[39] In particular it deals with the question how humans have changed their environment, which dates back to the seventeenth century, when Europeans became aware of the environmental devastation inflicted by their overseas expansion of

---

36 The correlation of economic growth and environmental degradation is a main interest in the research fields of *Environmental Economics* and *Ecological Economics*. The first discipline examines the effects of local or national environmental policies and focuses on empirical research of different pollution factors. The second approach using *Ecological Economy* depends on the ecosystem, with the need to preserve the natural capital and with emphasis on sustainability. It further refers to interdisciplinary and transdisciplinary fields of academic research, which claims to approach the correlation of the evolution of human economies within the ecological boundaries. For further information, see:
Van den Bergh, Jeroen C. J. M. "Ecological economics: themes, approaches, and differences with environmental economics." *Regional Environmental Change* 2, no. 1 (2001): 13–23. Accessed January 11, 2015.
Xepapadeas, Anastasios. "Ecological Economics." In *The New Palgrave Dictionary of Economics*. Edited by Steven N. Durlauf and Lawrence E. Blume, 599–604. Basingstoke: Nature Publishing Group, 2008.
37 Johnson, D. L., S. H. Ambrose, T. J. Bassett, M. L. Bowen, D. E. Crummey, J. S. Isaacson, D. N. Johnson, P. Lamb, M. Saul, and A. E. Winter-Nelson. "Meanings of Environmental Terms." *Journal of Environment Quality* 26, no. 3 (1997): 581.
38 Ibid.:581 http://www.oxforddictionaries.com/definition/english/nature.
39 Goudie, Andrew. *The human impact on the natural environment: Past, present and future*. Seventh ed. Chichester, UK: Wiley-Blackwell, 2013:3–7.

colonization.⁴⁰ Hence, environmental changes often imply that man *as polluter* plays the principal role. However, contrary to this, environmental changes are in generally naturally occurring in the larger scheme of global environmental changes and often used synonymously with the use of "global change".⁴¹

Global environmental changes can be divided in two aspects: *systematic* and *cumulative*.⁴² Systematic environmental changes occur in globally operating systems, like climate changes, whereas the cumulative changes are defined as deforestation, erosion and degradation. The assumption is that the systematic changes occur due to natural forces like volcanic eruptions or earthquakes, whereas the cumulative changes of environment are taking place throughout human history and have been introduced by mankind, like deforestations.⁴³

Since the seventeenth century, global environmental changes have been driven by humankind and modern achievements which exploit nature's richness. Environmental damage, as it occurs today, is linked with human activities and their increasing efforts to reach wellbeing and security.⁴⁴ To improve living standards, humans must, of course, exploit natural resources to a certain degree. As demonstrated in Inner Mongolia, coal is used as a source of energy and the mining industry involved has been destroying the natural environment.⁴⁵ On the one hand, the exploitation of natural resources has supported the rapid development of Chinese society, but on the other hand, the remarkable growth of China has become a synonymous with the environmental damage. In China, the degree of environmental degradation appears to increase in equal measure

---

40 Ibid.:3
41 Roy E. Cordato. "The Polluter Pays Principle: A proper Guide for Environmental Policy." *Institute for Research on the Economics of Taxation Studies in Social Cost, Regulation, and the Environment*, no. 6 (2001): 1–21. Accessed July 16, 2015.
Tobey, James A., and Henri Smets. "The Polluter-Pays Principle in the Context of Agriculture and the Environment." *The World Economy* 19, no. 1 (1996): 63–87.
42 Jiang, Hong. *The Ordos Plateau of China: An endangered environment*. UNU studies on critical environmental regions 1035. Tokyo: United Nations University Press, 1999:1.
43 Ibid.:1
44 Goudie, "The human impact", 27–69.
45 Xu Kangning, and Wang Jian. "An Empirical Study of A Linkage Between Natural Resource Abundance and Economic Development." *Economic Research Journal* 1 (2006). Accessed July 16, 2015.
Wen, Zhao. "An Empirical Study of The Linkage Between Resources Development And Economic Development-Taken Shanxi Province as example." *Energy Procedia* 5 (2011): 1394–98.

with the economic development.[46] China's current environmental problems are often linked to the rapid economic growth during the last decades. But economic growth is essential for sustainable development as it helps to alleviate poverty and increase the per capita income.[47] For any human society, prosperity is the basis for the development of environmentally aware thinking. This will be discussed in the following section.

1.4.1 Economic Growth and Poverty Alleviation

To evaluate the degree of environmental destruction and its interdependence with economic growth, two parameters are of importance: income and respective wealth. Income distribution is highly unequal in China.[48] The effect of economic growth on poverty alleviation has been part of several studies supported by the OECD and World Bank.[49] It has thus been found that economic growth is a reliable tool to reduce poverty and increase the living conditions in developing countries.[50] The increase of the GDP is considered as a "pro-poor" occurrence meaning that the population's poverty tends to fall with economic growth of a country. China has successfully followed a path of poverty reduction since 1980,

---

46 Wen Chen. "Economic Growth and the Environment in China: An Empirical Test of the Environmental Kuznets Curve Using Provincial Panel Data." Unpublished manuscript, last modified July 16, 2015. Available at: http://www.policyinnovations.org/ideas/policy_library/data/01447/_res/id=sa_File1/paper.pdf.
47 Lonnie K. Stevans and David N. Sessions. "The Relationship Between Poverty and Economic Growth Revisited." *Journal of Income Distribution* 17, no. 1 (2008): 5–20. Accessed July 16, 2015.
48 Griffin, Keith B. and Zhao Renwei, Eds. *The distribution of income in China*. Basingstoke, Hampshire: Macmillan, 1993.
    Riskin, Carl, Renwei Renwei, and Shi Li, eds. *China's retreat from equality: Income distribution and economic transition*. Asia and the Pacific. Armonk, NY: Sharpe, 2001.
    Li, Shi, Hiroshi Satō, and Terry Sicular. *Rising Inequality in China: Challenges to a Harmonious Society*. Cambridge: Cambridge University Press, 2013. Available at: http://site.ebrary.com/lib/alltitles/docDetail.action?docID=10729907.
49 Stephan Klasen. "Economic Growth and Poverty Reduction: Measurement and Policy issues: OECD Development Centre." Unpublished manuscript, last modified December 15, 2014. Working Paper No. 246.
50 Nallari, Raj, and Breda Griffith. *Understanding growth and poverty // Economic Policies for Growth and Poverty Reduction: Theory, policy, and empirics*. Washington, D.C.: World Bank, 2011.

while constantly increasing its economic growth.[51] Although China sustains its economic growth in a socialist-market economy, the theory of trade liberalization and open economies has successfully lifted millions of people out of poverty.[52] Despite trade reforms not demonstrating a visible short term effect on poverty reduction, over longer periods it has been demonstrated that reforms in investment and technology will increase growth and the poor will benefit.[53] Ravallion and Chen (2004) suggested that periods of greater economic liberalization in China did not synchronize with times of reduced poverty. Instead, they concluded that as long-term perspective productivity increases, income-levels will rise, both with a positive effect on economic growth and poverty alleviation.[54] On average the rural poor account for sixty-three percent of people living in poverty worldwide. In China alone, the amount of rural poverty accounts ninety percent of the population.[55] Indeed, as Ravallion and Chen explain, "*Among the rural poor, those who are landless suffer more, than landholders.*"[56] This is largely because the rural poor depend on income from agriculture, fishing, forestry and small-scale industry. Economic growth, especially for those sectors, has largely contributed to poverty reduction, even when landholdings are small and technology is outdated.[57] Given the assumption that agriculture contributes to a great extent to economic growth, effective agricultural development policies are necessary to alleviate rural poverty in China.

The Twelfth Five-Year Plan, targeted the reduction of the rural poverty by continuously increasing productivity and economic growth. China's government has been working to ensure that economy, environment and social disparity are balanced. Effective assets to alleviate poverty in rural areas include:

1. Guaranteed access to water
2. Tenurial rights and capital to participate in agricultural markets
3. Security and ability to compete in the rural economy

---

51 Martin Ravallion. "How long will it take to lift one billion people out of poverty?" Unpublished manuscript, last modified December 11, 2014. Policy Research Working Paper.
52 Nallari and Breda "Understanding Growth and Poverty", 191, 194.
53 Nallari and Breda "Understanding Growth and Poverty", 194.
54 Ravallion, M., and S. Chen. "How Have the World's Poor Fared since the Early 1980s?" *Policy Research Working Paper 3341, World Bank, Washington, DC.* 2004.
55 Nallari and Breda "Understanding Growth and Poverty", 299.
56 Ibid.:299
57 Ibid.:299

As China's agricultural economy is mainly based on smallholders, a great challenge is to move these farmers out of poverty. Governmental policies support a sustainable development via the liberalization of markets and improving price incentives for smallholders. Establishing a domestic market, improving staple food crops and enhancing agricultural export, as well as stimulating the domestic consumption for high-value items are major contributors for the alleviation of poverty. Furthermore, to ensure a well-functioning market economy it is necessary to promote the farmers' access to financial services with a simultaneous reduction of uninsured risk exposure. Another tool to establish a balanced economy is the implementation of producer organizations and science and technological innovation.[58]

Using the proper incentives and investments in agriculture, the environment can be protected and natural resources harnessed to sustain watersheds and biodiversity. Reflecting the interdependency of agriculture and environment it should be of high priority of any society to make agriculture more sustainable, as agriculture is no longer prosperous without better stewardship of the natural resources. Most of China's rural environmental problems are linked to agricultural intensification and animal husbandry.

1.4.2 The Environmental Kuznets Curve

One hypothesis dealing with environmental degradation and economic growth is the Environmental Kuznets Curve (EKC). It dates back to 1955 and derives its name from Simon Smith Kuznets (1901–1985), who postulated in 1946 that income inequality first increases with economic development and will at some later point, fall (*Kuznets Curve*).[59] The EKC assumes that environmental deterioration accelerates in the early stages of economic development and improves with later stages of economic growth. Environmental damage and pressure on natural resources first increases with rising income and later declines, according to GDP growth.[60] Meaning that at levels of economic development comparable to subsistence levels, the impacts on environment and degradation are relatively

---

58 Nallari and Breda "Understanding Growth and Poverty", 309.
59 Simon Kuznets. "Economic Growth and Income Inequality." *The American Economic Review* XLV, no. 1 (March 1955). Accessed January 27, 2016. https://www.aeaweb.org/aer/top20/45.1.1-28.pdf.
60 Stern, David I., Michael S. Common, and Edward B. Barbier. "Economic growth and environmental degradation: The environmental Kuznets curve and sustainable development." *World Development* 24, no. 7 (1996): 1151–60. Accessed December 8, 2014. doi:10.1016/0305-750X(96)00032-0.

low and limited to biodegradable pollution. However with the intensification of agriculture and the transition to an industrialized economy, the resource extraction and non-biodegradable waste accelerates and the environment deteriorates to a higher degree. The watershed of the EKC is reached when environmental damage decreases and the economy changes to service-oriented industries and environmental awareness develops. This development is the basis for environmental policies and regulations, as well as renewable energies, which may result in a gradual decline of environmental degradation, if it occurs.[61]

Fig. 2: Environmental Degradation Kuznets Curve

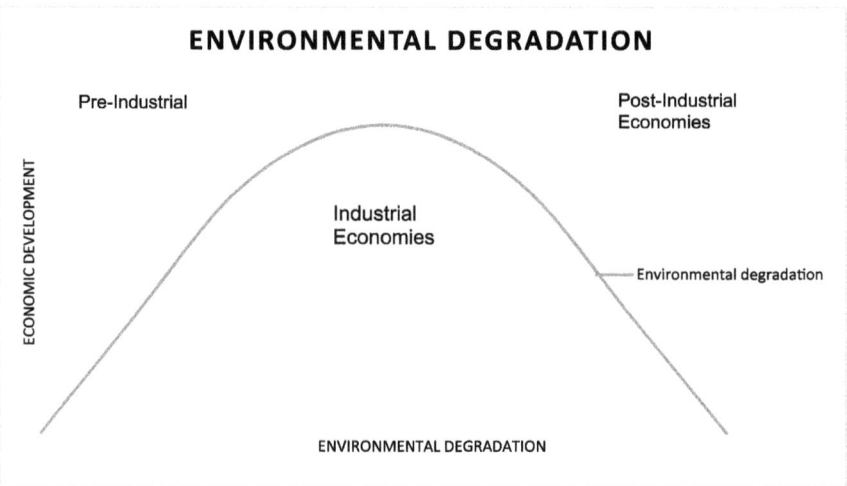

Source: Cilia Neumann (2016), *EKC*

Looking at the vast development of China and Inner Mongolia in particular, the environmental problems seem endless and impossible to address. Nevertheless it must be kept in mind that the "emerging" countries look to their neighbors and often environmental policies can be implemented within a short period. Today

---

Stern, David I. "The Rise and Fall of the Environmental Kuznets Curve." *World Development* 32, no. 8 (2004): 1419–39. Accessed December 8, 2014. doi:10.1016/j.worlddev.2004.03.004.
61 Dinda, Soumyananda. "Environmental Kuznets Curve Hypothesis: A Survey." *Ecological Economics* 49, no. 4 (2004): 431–55. Accessed December 8, 2014. doi:10.1016/j.ecolecon.2004.02.011.

the transformation from "old school" energy supplies to more modern sustainable energy-use through the adoption of new technologies and innovations is no longer a privilege exclusive to "developed" countries. For example, China has been actively supporting the renewable energies and in fact, the world's largest wind farms are in Inner Mongolia and Gansu.

Where environmental degradation is no longer increasing in relation to economic growth and increasing per capita income, the assumption is that the emerging and developing countries may reach the EKC turning point, more quickly and at a lower level of environmental degradation. At a certain point, depending on an average per capita income, a society may starts to reflect upon natural resource extraction and sustainability, non-biodegradable waste and pollutants and their impact on ecosystems.[62] In the case of China, Shen Junyi (2006) has proven that the assumed EKC relation exists in COD, Arsenic and Cadmium emissions, still the poorer provinces have to increase their wealth, before they start to react and improve their environment.[63] This circumstance is of relevance, when examining the development of environmental improvement in Inner Mongolia, as the income distribution is still highly unequal, even though Inner Mongolia has remarkably improved its GDP ranking.[64]

By adopting environmental policies and implementing "green" and renewable technologies to reduce emissions, water and air pollution, as well as ceasing the predatory exploitation of nature at an earlier stage, a "breakthrough" in the theoretical construct could be possible (see figure 2). The effect will be a "tunnel" through the EKC, where economic development is still accelerating, but the environmental damage is no longer increasing at the same rate.[65] This scenario indeed would be a milestone for all countries still in a developing process and dealing with environmental deterioration.

---

62 Shen, Junyi. "A simultaneous estimation of Environmental Kuznets Curve: Evidence from China." *China Economic Review* 17, no. 4 (2006): 383–94. Accessed July 16, 2015. doi:10.1016/j.chieco.2006.03.002.
63 Ibid.:393
64 Within the domestic ranking it has moved from 17 in 1978 to 6 in 2014. For detailed information the author refers to the National Bureau of Statistics of China (NBS): http://data.stats.gov.cn/english/swf.htm?m=turnto&id=3.
65 Dinda, "Environmental Kuznets Curve", 445.

Fig. 3: Environmental Degradation Tunnel Effect Kuznets Curve

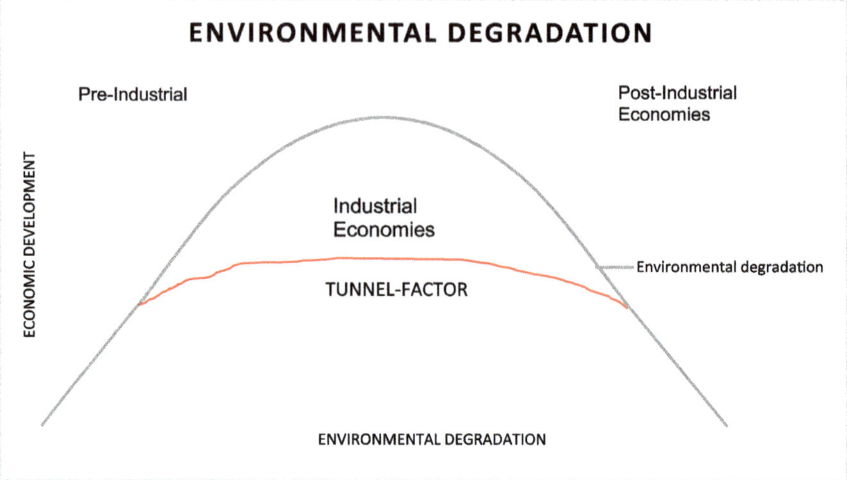

ENVIRONMENTAL DEGRADATION

Pre-Industrial — Industrial Economies — Post-Industrial Economies

ECONOMIC DEVELOPMENT

TUNNEL-FACTOR — Environmental degradation

ENVIRONMENTAL DEGRADATION

Source: Cilia Neumann (2016), *Tunnel-Effect of the EKC*

The tunnel-effect of the EKC assumes that the country under investigation is fully in the transformation process between pre-industrial and post-industrial economies, and that it has a well-educated, wealthy society.[66]

China yet hast to develop the necessary well-educated, wealthy society and thus could be a serious threat to environmental sustainability and economic growth without the destruction of natural resources because sustainability requires governmental stewardship. China's government must play the central role in the process of a sustainable development with civil society taking a supportive role of these developments. As Yang, Niu and Xu (2011) argue, the government must develop policy decisions, mobilize basic education in rural areas and take control as an inspector, supervisor and organizer of environmental protection.[67] China's Vice Premier, Pei Yan Zeng, mentioned in 2005 during the twenty-third Session of the United Nations Environment Program Governing Council

---

66 Munasinghe, Mohan. "Is environmental degradation an inevitable consequence of economic growth: Tunneling through the environmental Kuznets curve?" *Ecological Economics* 29, no. 1 (1999): 89–109. doi:10.1016/S0921-8009(98)00062-7.
67 Lijing, Yang, Niu Yonghong, and Xu Yanli. "Sustainable Development and Formation of Harmonious Nature." *Energy Procedia* 5 (2011): 629–32. Accessed June 11, 2013. doi:10.1016/j.egypro.2011.03.110.

meeting that, "*in the aspect of environmental protection, the Government will focus on five measures taken to effectively protect the environment.*"[68] These measures include establishing clean production, building a conservation-oriented society, implementing pollution prevention and control, supporting ecological quality and protecting the environment.[69] If this is feasible, effective environmental policies as the environmental law of the polluter pays principle are basic approaches for the turning point of the EKC in China.

Civil society is the secondary actor in these developments. As a participator in the debate, the public acts as consumer and producer of goods and services. This will affect the environment in diverse ways as the people have a say in the environmental discussion. They are an active part of the transformation of environmental protection changing from the government over enterprises to its citizens.[70] The successful combination of both actors, governance and society, is a chance to reach the turning point of environmental destruction at an earlier date, even though an equality in income-distribution has not been achieved yet and socio-economic disparity still plays a significant role in the developing country.

1.4.3 Developing Environmental Awareness

Early models from the 1970s assumed that education is the key to pro-environmental behaviour.[71] But the assumption that the linear progression of environmental knowledge, leading to environmental attitudes, which turn into pro-environmental behaviour, has been proven to be wrong.[72] Nevertheless, most environmental-based NGOs still rely on older research suggesting that campaigns based on knowledge transfer will lead to more sophisticated behaviour. However, it has been demonstrated that there is a difference between attitude and behaviour[73], and several studies have been conducted attempting to explain that gap.[74] In this section, the author will first address three linear models of Fishbein and Ajzen (1975), Rajecki (1982), Hines, Hungerford and

---

68 Zeng Peiyan. "Strengthen environmental protection and achieving sustainable development." Nairobi, February 21, 2005. Accessed January 27, 2016. http://www.unep.org/gc/gc23/documents/Zeng_Peiyan_speech.pdf.
69 Lijing et al., "Sustainable Development and Formation", 4.
70 Ibid.:4
71 Kollmuss and Agyeman. "Mind the Gap", 239–60.
72 Ibid.:241
73 Ibid.:242uu.
74 For detailed information on different models of developing environmental behaviour the author refers to the works of:

Tomera (1986) and finally present the models of Fietkau and Kessel (1981) and Blake (1999), which are more differentiated in terms of factors influencing pro-environmental behaviour.

The first model of Fishbein and Ajzen has become the most influential attitude-behaviour model in social psychology because unlike other models, it can be used to calculate measurable behavioural outcomes rather than relying solely on interpretation.[75] Fishbein and Ajzen state that attitudes do not determine behaviour directly, but they do have a strong influence on the individual's actions. Instead, the way the individual's performance is judged by others plays a significant role in pro-environmental behaviour. Fishbein and Ajzen distinguish between rational behaviour, which is based on the systematic use of information, persuading the individual to act pro-environmentally and normative pressure, provided by others, which influences a person's decision to behave pro-environmentally.

Tab. 1: Fishbein & Ajzen 1975

Fishbein & Ajzen 1975[76]	Content
Rational Behaviour	Systematic use of knowledge and information influence behavior
Normative Pressure	Concerns about behavioural prescription of others

Another model that can be used to evaluate potential behaviours is based on the linear progression, assuming that environmental-issue knowledge will lead to engagement in pro-environmental behaviour. Rajecki attempted to explain this gap between attitude and behaviour and he suggests that the individual's involvement in environmental issues, either personal experiences or societal norms, plays a significant role in whether they will act pro-environmentally.[77]

Newhouse, Nancy. "Implications of Attitude and Behaviour Research for Environmental Conservation." *The Journal of Environmental Education* 22, no. 1 (1991): 26–32. Accessed April 22, 2015.
75 Kollmuss and Agyeman. "Mind the Gap", 243.
76 Fishbein, Martin, and Icek Ajzen. *Belief, attitude, intention, and behavior: An introduction to theory and research.* Addison-Wesley series in social psychology. Reading, Mass.: Addison-Wesley Pub. Co; Addison-Wesley, 1975.
77 Kollmuss and Agyeman. "Mind the Gap", 242.

Tab. 2: Rajecki 1982

Rajecki 1982[78]	Content
Direct vs. Indirect Experience	Stronger influence of personal experiences (direct) affect pro-environmental behavior
Normative Influences	Social norms, traditions & customs form attitudes
Temporal Discrepancy	Attitudes change over the time
Attitude Behaviour Measurement	Discrepancies in attitudes and actions (Newhouse, N. 1991) influence pro-environmental behaviour

The third model, selected to explain the correlation of attitude (awareness) and action (engagement) is the framework outlined by Hines, Hungerford and Tomera. Although the framework of Hines, Hungerford and Tomera is more detailed, the given factors do not fully explain pro-environmental behaviour. Whether a person engages in environmental issues or not depends on several situational factors like economic constraints, social pressures, personal opportunities and obstacles, all of which influence a person's decision.[79]

Tab. 3: Hines, Hungerford & Tomera 1986

Hines, Hungerford & Tomera 1986[80]	Content
Knowledge of Issues	Person has to be familiar with causes of environmental problems
Knowledge of Action Strategies	Knowledge about the ways to minimize the environmental problems
Locus of Control (Rotter 1966)	The individuals believe they control the events that effect (?) them
Attitudes	pro-environmental attitudes lead to pro-environmental behavior
Verbal Commitment	Verbal commitments lead to engagement in pro-environmental behaviour
Individual Sense of Responsibility	Greater sense of responsibility leads to higher engagement in pro-environmental behaviour

---

78   Rajecki, D. W. *Attitudes, themes and advances*. 1st ed. Sunderland, Mass.:Sinauer Associates, 1982.
79   Kollmuss and Agyeman. "Mind the Gap", 243.
80   Hines, Jody M., Harold R. Hungerford, and Audrey N. Tomera. "Analysis and Synthesis of Research on Responsible Environmental Behavior: A Meta-Analysis." *The Journal of Environmental Education* 18, no. 2 (1987): 1–8. doi:10.1080/00958964.1987.9943482.

The commonalties between these three models are that they all follow a linear progression approach, considering that a certain achievement (education/knowledge, commitment/responsibility, attitude/external perception) inevitably lead to a proper conduct. The important differences in interpretation are that Fishbein and Ajzen are not relying on education as crucial to establish pro-environmental behaviour, whereas Rajecki, Hines, Hungerford & Tomera explicitly refer to education and knowledge as the opinion-forming and decision-making factors to develop environmental awareness.

Two other models should be mentioned, as they are more differentiated in their predictions of environmental behaviour. Fietkau and Kessel and Blake use sociological and psychological factors as well as barriers, which influence pro-environmental behaviour or the lack of it.[81]

Fietkau and Kessel refer their model to five variables that influence either directly or indirectly a person's pro-environmental behaviour. These variables underlie independent factors, which are changeable.

Environmental attitudes and values reflect people's environmental perception and are internal factors forming the personal decision to act pro-environmental or not.

Possibilities to act environmentally are external factors, like infrastructural or economic constraints, hindering pro-environmental behaviour.

Environmental knowledge determines a person's attitudes and values. [I state that it is both an internal and external factor, as it enables environmental concern (internal) and the absence of awareness, when not educated (external).]

Incentives for pro-environmental behaviour are also internal factors, which influence a person's decision to act pro-environmental, when the outcome of this behaviour is of personal advantage (increase of life quality).

---

81 Fietkau, Hans-Joachim and Hans Kessel, eds. *Umweltlernen: Veränderungsmöglichkeiten des Umweltbewußtseins; Modelle, Erfahrungen*. Schriften des Wissenschaftszentrums Berlin Sozialwissenschaft und Praxis. Internationales Institut für Umwelt und Gesellschaft 18. Königstein/Ts.:Hain, 1981.
Hellbrück, Jürgen, and Elisabeth Kals. *Umweltpsychologie*. Basiswissen Psychologie. Wiesbaden: VS Verlag für Sozialwissenschaften, 2012. Accessed June 18, 2015. doi:10.1007/978-3-531-93246-0.
Blake, James. "Overcoming the 'value-action gap' in environmental policy: Tensions between national policy and local experience." *Local Environment* 4, no. 3 (1999): 257–78. doi:10.1080/13549839908725599.

Perceived consequences of pro-environmental behaviour are external factors, as the feedback can be either intrinsic (satisfaction through "doing the right thing") or extrinsic (not littering, as it is perceived as dishonorable).

Tab. 4: Fietkau & Kessel 1981

Fietkau & Kessel 1981[82]	Content
Internal factors influencing pro-environmental behaviour	Attitudes and values, Behavioural Incentives
External factors influencing pro-environmental behaviour	Possibilities to act ecologically (infrastructure, economy) perceived feedback about environmental behaviour
Environmental Knowledge	Acts as a determinant for attitudes and values

Overall, the model of Fietkau and Kessel is more methodically sound than the early models of Rajecki, as it includes more factors, which influence and help to make the final decision to act pro-environmental. Still the question what shapes pro-environmental behaviour is such complex that it is nearly impossible to visualize all of the interactive factors, which influence pro-environmental decision-making.

Finally, Blake uses the term *Value-Action-Gap* to describe the differences between attitude and behaviour.[83] He actually criticizes earlier models, because of their lack of individual, institutional and social constraints and the assumption that people intentionally make systematic use of available knowledge.[84] But theory shows that pro-environmental acting is non-exclusively based on available information of environmental issues, rather responsible action is based on several factors, which form a subjective picture of environmental concern. Blake argues that the eminently rational appeal to act ecologically fails, because people's values are negotiable, transitory and contradictory.[85] This means that even a strong individual environmental concern and awareness can be outweighed by other conflicting interests. For example, the decision to take the car, when it is heavy raining overrides the personal value of riding the eco-friendly bicycle. This decision stands for the controversial internal factors, which stimulates a person's decision.

---

82 Fietkau and Kessel "Umweltlernen".
83 Kollmuss and Agyeman. "Mind the Gap", 246.
84 Ibid.:246
85 Redclift, Michael R. and Ted Benton, eds. *Social theory and the global environment*. Repr. Global environmental change series. London: Routledge, 1997.

Institutional rules or governmental policies are external factor, which might convince the individual to act pro-environmental or not. If the person lacks trust in the local or national government, the point of *Responsibility* is maybe more crucial for the decision-making process than the *Individuality*. The third barrier of Blake is *Practicality*, which hinders pro-environmental behaviour, because lack of time, money or information counterbalance attitudes or intentions.[86]

Although the model of Blake takes several factors into account, he obviously forgoes the effect of social factors (cultural norms, normative pressure), which were mentioned by Fishbein and Ajzen to have a strong influence on behaviour. Nevertheless, the communalities of the two models by Fietkau and Kessel as well as Blake are that they differentiate between internal and external behavioural incentives, which form a person's decision.

Tab. 5: Blake 1999

Blake 1999[87]	Content
Three barriers to act pro-environmental	1. Individuality (lack of interest, laziness) 2. Responsibility (lack of trust, lack of efficiency, no property) 3. Practicality (Lack of time, money, information, facilities, encouragement etc.)
External Barrier	Social / Institutional
Internal Barrier	Individual
Environmental Concern vs. Pro-Environmental Behaviour	The concern has to pass the three barriers to act pro-environmentally

Relevant for this research and the focus on Inner Mongolia are in particular the models, which include different influential factors, even though they are to some extent arbitrary. Therefore my approach is that the author tries to distinguish between demographic factors (Age, sex, education, level of income, ethnicity etc.), external factors (institutional, social, cultural and environmental) and internal factors (motivation, awareness, environmental knowledge, values, attitudes, responsibility etc.).

Examining the occurrence of environmental awareness, we should start from the premise that environmental knowledge exists and that any kind of emotionally involvement is necessary to establish environmental awareness and attitude.

---

86 Kollmuss and Agyeman. "Mind the Gap", 247.
87 Blake, "Overcoming the 'value-action gap'".

2. Research Framework

This thesis, which is dealing with the idea of sustainable development in Inner Mongolia, uses an anthropological approach to look at the environmental changes in this region from different, often competing human and institutional influences, in the context of environmental awareness. It will examine the economic, social, and political limitations which hamper the balance between the ecological, social, and economic needs of the society. Usually development is driven by a particular need, in China it is economic growth; in Inner Mongolia it is the ability to guarantee an average per capita income to lift its rural population out of poverty. When concentrating on a certain need without considering long-term future impacts, further environmental damage is a negative result typical of such an approach. Therefore, the broader variables like demographics as well as internal, and external factors[88] in Inner Mongolia must be located based upon their involvement within this context.

2.1 Project Setting

The thesis is embedded as the second study within the larger project *Water: Chance and Danger – Environmental Strategies along the Yellow River (1820–1970)* subsidized by the Deutsche Forschungsgemeinschaft (*German Research Foundation*). The first study of this same project, "Change of the Natural Environment: Immigration Society in Hetao from 1850s to the 1950s," analyses the impact of Han-Chinese in-migration into Inner Mongolia, specifically into the Hetao Area situated in the upper reaches of the Yellow River in north-west China. Both dissertations focus on Inner Mongolia and address the human impact on the environment with an emphasis on agriculture and animal husbandry as these are the basis for typical subsistence economies of the locally dominating ethnic groups.

This study focuses on economic and environmental issues as well as on social aspects related to sustainability in agriculture and animal husbandry in Inner Mongolia. The author argues on the basis of the Environmental Kuznets Curve that an average increase of per capita income and a rise of economic growth would improve the agriculture and environmental situation of Inner Mongolia dramatically. Thus economic growth is a secure factor for optimizing farmers' low incomes in terms of productivity and cost efficiency, as well as resilience

---

88 See page 35–37 for a more specific list of these factors.

and self-sufficiency. The *Environmental Kuznets Curve* confirms the positive repercussions of economic growth on environmental damage. But the successful implementation of environmentally friendly and sustainable technologies depends on the farmers' awareness of a healthy ecology. Therefore reflections upon sustainability and perceptions of environmental degradation are in the focus.

2.2 Field Research

The first quantitative survey with thirty potato-crop farmers was conducted during September 2013 in the intensive agricultural zone of Siziwang Qi Banner in Inner Mongolia. A second survey of 268 randomly chosen informants was distributed in various banners (qi 旗), leagues (meng 盟), prefecture-level cities (dìjí shì 地级市), one city-district (qù 区) and one county-level-city (xiànjí shì 县级市). Three individual face-to-face interviews, based on the method of a semi-structured interview to allow new questions and give the informants the possibility to provide more information on specific issues were conducted in September 2013. In these interviews three different business owners were interviewed to illustrate typical rural economies and business models and the diversity of ethnicities in Inner Mongolia. An additional goal of the interviews was to eliminate the prejudice connected to ethnicity and profession: a Mongolian potato seed breeder, a Han-Chinese livestock breeder and a taxi driver of Hui nationality.

All questionnaires were divided into three parts in order to a more organized format for a comparison of the results:

The first part is a specific household survey using open-ended questions, which in addition to general information (*age, sex, ethnicity, education, etc.*), emphasizes the social structure of the household (*family members and education, occupation of head of household, future perspectives for the children etc.*). These societal factors might influence the development of environmental awareness and later the development of pro-environmental behaviour and were thus included in the survey.[89] Furthermore, this information provided important background information and served to connect the individual interviews to the wider array of surveys.

The second part of the survey was addressed specifically to the head of the household and asked questions about agro-economic issues related to his or her agricultural production system (*acreage/pastures, crops/animals, yield, farm*

---

[89] Blake, James. "Overcoming the 'value-action gap' in environmental policy: Tensions between national policy and local experience." *Local Environment* 4, no. 3 (1999): 257–78. doi:10.1080/13549839908725599.

subsidies, re-investment and income). This was particularly important for the potato-related fieldwork because the surveys confirmed that while potato production is a matter of the national economic growth policy, it mainly involves smallholders who are struggling with the economic and environmental conditions of Inner Mongolia's dryland agriculture. To gain a deeper understanding of the threats facing herders, regarding grassland degradation and livestock increases, the focus on agro-economics is especially interesting as it helps to reveal the difficulties of remaining economically independent *and* at the same time sustainable, a precarious balancing act for the majority of rural population in Inner Mongolia.

The third section of the survey attempts to reveal individual farmers' and herdsmen's ideas about environmental awareness, here specifically understood as an *environmental protection awareness* and the interviewees' stances towards social and environmental matters. This section is the most important for this dissertation as it deals with environmental destruction and issues affecting the business of the local population. Through these questions it was possible to better understand and evaluate how environmentalism has developed in the region and the ecological sensibility of the interviewees. One particular point of focus was whether reflections on environmental problems led to more environmental awareness and sustainable action by individuals, regardless of their level of political engagement.

One of the major challenges of this section was to define environmental awareness in terms of the idea of environmental protection and environmentalism. As long as the participants mentioned protective measures (Category: *Prevention Methods for Environmental Degradation*) as well as root causes for environmental problems (Category: *Reasons for environmental problems*) it was fair to assume that these individuals anticipated how their environment would change with respect to environmental issues that would affect them. However the question was raised as to how to categorize participants who did not provide any answers to these specific questions. It would be incorrect to assume that a person, who for whatever reason did not provide answers in this section, does not possess any environmental awareness. A blank interview sheet only reveals that the interviewee did not answer the question. However, if the interviewee answered "I don't know" or "I do nothing," this is much more revealing. "Doing nothing" can be a conscious decision, despite the interviewee having had reflected on causes of environmental problems. For example, it could be that this answer was triggered by external factors limiting the ability of the person to act. Although "I don't know" can be the answer regarding root causes of environmental degradation, it was often the case that the person was still able to name preventative measures to environmental methods and astonishingly implement them. With this complex structure in mind, analyzing the narratives was only possible by looking at the provided information in its overall context and looking at the different codes[90], which might imply a deeper environmental understanding of the participant.

2.2.1 Challenges

Difficulties regarding the survey arose shortly after discussing the framework of the questions. Research on environmental problems in China must be handled with sensitivity due to censorship therefore the surveys were designed to balance the core theme of environmental degradation in relation to agro-economic issues facing Inner Mongolian farmers or herdsmen. Both the agricultural and husbandry sectors face challenges of global economic competitiveness, which influence the handling of environmental issues.

Without personal engagement and relying on good relationships with local people, the fieldwork would have been very difficult. Especially due to the sensitivity of the topic –environmental degradation – which still must be handled cautiously and has only slowly become a matter of public interest in the last decade. Thus the people in Inner Mongolia are reserved and critical when discussing the topic with strangers. It is essential to understand that Chinese society relies on relationships called *guānxi* (关系).

As an anthropologist, the importance of support from a local, who helps to establish connections with other people, was viewed as essential in order to gain contact with the informants. With the support of an Inner Mongolian dealer in potato seeds as well as contact with employees of a Swiss agro-business company, the author was able to speak to local potato farmers and receive information from government circles, including an interview with the Head and Deputy Head of the Agricultural Bureau in Hohhot. Due to the sensitive, business-related nature of some of the information gained in these interviews it was necessary that it be handled confidentially.

2.2.2 Questionnaire

The questionnaire and supplementary information were both provided to the participants in Chinese. The recorded answers were handwritten by the participants. The following figure illustrates the structure of the questionnaire and the composition of questions into three categories.

---

90 The structure of the narrative, including categories and codes to analyze and evaluate the provided information is subject of chapter 2.2.2.

Tab. 6: Structure of Questionnaire

A. Personal Information	B. Land-use practices	C. Environmentalism
Age	Total area of land?	Fear of environmental degradation?
Sex	Total area of cultivated land? Grassland?	What kind of degradation?
Ethnicity	What kind of crops? Animals?	Drought?
Languages	Decision for choosing these crops/animals?	Salinization?
How long have you lived in your area?	Since when do you cultivate these crops? Raise these animals?	Soil erosion?
Occupation	Crop rotation? Rotational grazing?	Grassland degradation?
How many people live in your household?	Annual household subsidy?	Others?
Children	Annual household net income?	Reason for environmental degradation?
Children's education	How much money for re-investment?	Prevention of further degradation?
Future perspective for your children?	What kind of re-investment?	
Are you Head of the household?		

Source: Cilia Neumann (2016), *Questionnaire*

Part A. was designed to evaluate on a comparative basis due to the fact that certain factors influence different attitudes and opinions towards environment. Therefore, ethnic background, age, education, economic sector and income distribution were chosen as survey criteria in order to help distinguish similarities as well as disparities between the participants. With reference to education, later statements from participants on environmentalism could be connected with theories on development of environmental awareness and pro-environmental behaviour.

Part B. considered the individual economic development of each household. The *Annual Household Net Income* (AHNI) is a guideline for the EKC to show the correlation between income and environmental awareness. This second section of the survey provided important information on the economic situation of each household and allowed for the analysis of the cropping patterns of Inner Mongolian smallholders as well as different methods of livestock production.

Part C. dealt with the most the most sensitive issue of the survey: Unfortunately due to the political sensitivity of the topic, the questions had to be formulated very generally. Despite this setback and thanks in large part to personal contact with the informants, detailed personal answers on environmentalism could still be obtained. Precise distinctions were made regarding the diverse environmental problems affecting both farmers and herdsman in Inner Mongolia. Climate conditions were also mentioned, namely drought and the specific condition of the Mongolian *dzud,* as well as the phenomenon of desertification, siltation and different pollutants. The main focus of this section was on the aspect of "fear". Specifically, do the people in Inner Mongolia fear environmental degradation and what do they think is responsible for the damage? The question, whether the local community supports sustainable policies and reflects upon prevention methods, lead to answers regarding the role of the government in the opinion–forming process.

Coding of the provided answers

To analyze the answers relating to the specific topic on environmental awareness it was essential to translate the answers into a code for categorization. Interviewees who provided no answers in all relevant questions with regard to environmental content could not be included in the evaluation as they would provide a false picture of the interpretation. But as long as the participant answered just one question on environmental issues, either stated fears or worries or even mentioned specific problems, it was assumed that the individuals had reflected upon their environment and scrutinized environmental problems. This was thus interpreted as possessing an idea of environmental protection awareness. Two examples were chosen to illustrate the complexity of analyzing and matching the answers: Interviewee No. 3[91], a forty-year old camel breeder (for tourism purpose) from Alashan prefecture stated that she has no fear of environmental degradation, either dzud or grassland loss. Nevertheless, as a camel breeder she did worry about the increase of grassland degradation and saw the root cause as heavy "overgrazing". Her engagement to prevent further degradation was to "reduce the animals". Interviewee No. 47[92] is a potato and sunflower farmer in

---

[91] Full interview of No. 3 accessible on attached CD.
[92] Full interview of No. 47 accessible on attached CD.

Wumeng, Siziwang Qi. He answered that he had "no" fear of environmental degradation and was not worrying about any of the specific environmental problems in Inner Mongolia (soil erosion, desertification, grassland degradation). Thus his answer that the "bad vegetation" is responsible for the degraded land was unexpected. He actively plants trees to prevent further increase of soil erosion. In either case interviewer prejudice would lead to a false interpretation if assuming that no answer implied denial or ignorance of environmental issues. In fact, several reasons could have contributed to a lack of answers including time pressure, little interest in the survey or reservations regarding the questions.

To circumvent the participants' bias regarding the sensitivity of the topic as well as to include the possibility of purposely providing false answers – both are issues the interviewer has to consider, when working with standardized questionnaires – the participants were introduced to the survey by an Inner Mongolian middleman who explained the topic and the background of the project. Thus, it could partly be ensured that only the eldest, the head of the household or the business owner answered the questions. Participants under thirty years of age and those who were not the head of the household were double-checked regarding their quality of the answers. Often it seemed that the children or grandchildren filled in the questionnaire on behalf of the interviewed candidate. Here it is necessary to acknowledge that a standardized questionnaire does involve the potential risk of not being very meaningful and methodically sound due to the inability for follow-up with individuals about their answers. However, cognizance of these challenges before distributing the survey and during its evaluation supported a thorough and thoughtful interpretation of the provided information, which despite the potential drawbacks was essential to this project.

Tab. 7: Coding and Categorizing

Answer	Code	Category
Fear of environmental degradation?	Sensitivity	Environmental Awareness
What kind of degradation fear most?	Knowledge/Observation	Environmental Awareness
Salinization	Knowledge/Observation	Environmental Awareness
Soil Erosion/Soil Acidification	Knowledge/Observation	Environmental Awareness
Water Pollution	Knowledge/Observation	Environmental Awareness
Grassland Degradation	Knowledge/Observation	Environmental Awareness

Answer	Code	Category
Air Pollution	Knowledge/Observation	Environmental Awareness
Others	Knowledge/Observation	Environmental Awareness
Reason for environmental degradation?	Knowledge/Observation	Environmental Awareness
Cultivation Methods	Reaction	Pro-environmental Behaviour
Livestock Methods	Reaction	Pro-environmental Behaviour
Prevention Methods	Engagement	Pro-environmental Behaviour

Category	Definition	Coding Principle
Environmental Awareness (E.A)	E. A. is given when the participants mentioned and or reflected on 1. Fear of specific environmental issues 2. Reasons/causes for environmental degradation	Definition 1. and/or 2. had to be given
Pro-Environmental Behaviour (P-E.B)	P-E.B is given, when the participant provided 1. Any information on personal engagement and/or improvement activities	Definition 1. had to be given

2.2.3 Realization and Distribution

The surveys were conducted separately within a period of three months starting in September 2013 in cooperation with additional field assistance. The individual semi-structured interviews were conducted personally by the author during September 2013. The interview with potato-growers in Siziwang Qi Banner was distributed in October 2013 by an employee of a Swiss agro-business company during field visits. Later the completed questionnaires were sent to the author, manually translated and archived in Excel.

The second broad questionnaire was distributed by a personal friend, Shi Li, Professor of Entomology at the Agricultural University of Inner Mongolia. These were conducted in November 2013. This second questionnaire was distributed through students of the Inner Mongolia Agricultural University of Hohhot (内蒙古农业大学). Several students were recruited and contacted before they travelled home to their families for winter holidays. The students needed to fulfil certain requirements to complete the questionnaire focusing on family income, which had to be either from farming or livestock. Furthermore, it was the head of the household who would be interviewed and not the student in order to guarantee a minimum level of maturity and experience when answering the questions. These surveys were returned at the end of a three month period. These were also manually translated and copied into Excel sheets for further evaluation.

Depending on the students' residence and the participants' origin, both narratives were able to reach a wide variety of people, and thus the answers are quite diverse. Han-Chinese as well as Mongols took part in the interviews. Some participants were also of the Hui minority. Economic background was also variable, and this applied to both farmers and herdsmen. Thus it was nearly impossible to draw inferences about ethnic origin as it relates to subsistence economy. Some of the participants were in the farming sector, while others lived fully as pastoralists.[93] Often additional income was guaranteed from tourism, for example, renting yurts or providing camels for touring.

---

93 Pastoralism is the economy concentrating on raising livestock. The term pastoralist can be used synonymously with herdsman and herder.

3. Region of Study

In this chapter the author traces the history of Inner Mongolia from Ming to present day, showing that over the time there were migration trends, which have led to today's ethnic blending of the Han and Mongolian community in Inner Mongolia. The historical development of this transition zone was to a certain extent responsible for the environmental problems facing today. China's immense and fast population growth on limited arable land led to early agricultural problems. Today the gap of supply and demand is tried to be solved through production and yield increase, which raise concerns with regard of ecological stress and calls for agricultural sustainability.

Throughout history, the Han-Mongolian community was characterized by periods of peaceful coexistence and aggressive expansion policy by alternating initiators. This chapter will start with the Han-Chinese expansion into the North since the Han dynasty and the following back and forth movement of the Mongolian tribes until Qing dynasty. This geopolitical development has marked the environmental changes of Inner Mongolia's landscapes and has to a great extent formed the scenery of grassland and farmland.

First the region of study will be shortly presented and then a more detailed historical development of Inner Mongolia will be outlined. What follows is a brief introduction in the economic and environmental background of China. Explicitly the geopolitical situation of Inner Mongolia with emphasis on different Inner Mongolian drylands and their issue of desert expansion and Hetao Irrigation District, which is marked by its unique feature of an artificial irrigated agriculture area will be illustrated in this chapter.

Map 1: Prefectures of Inner Mongolia

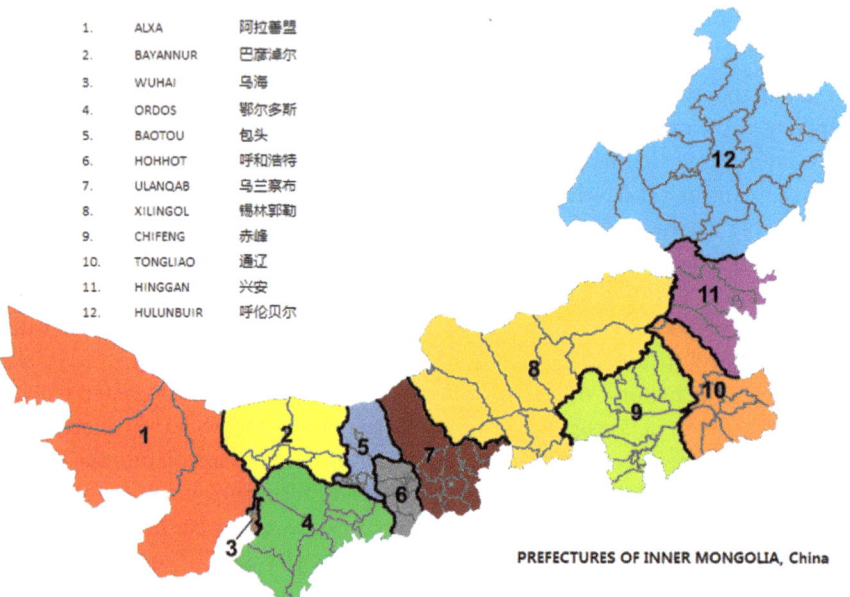

Source: Cilia Neumann (2016), *Inner Mongolia's Prefectures*

The region of study is the Autonomous Province of Inner Mongolia. Research for this project on agricultural development with an emphasis on the national potato policy has been conducted in areas where crop cultivation is the main source of income in order to guarantee homogenous answers. Thus the predominantly agricultural banner Siziwang, situated in Ulanqab prefecture, was chosen, as the majority of farmers there cultivate potatoes.

To help develop an average understanding of Inner Mongolians' perceptions regarding environmental change and damage, places of origin at first, played an inferior role. Thus, research was conducted in several areas, including different leagues and banners in Inner Mongolia, all dependent on the informants' hometown. Due to the quantity of questionnaires, answers from every prefecture could be obtained. Later for the evaluation on pro-environmental behaviour and the correlation of attitude and action, place of descent, income level and the distinction between livestock raising and farming became an important factor in the analysis of developing environmental awareness.

3.1 Historical Review

Three thousand years ago the region which comprises today's Inner Mongolia was the homeland of several different nomadic tribes, who would later be correlated with the Huns.[94] The recorded history of present day Inner Mongolia as political-geographical region began only with the conquest of Qing Dynasty (清朝 1644–1911). Before this time, references did not speak of Inner Mongolia as a region, and it was only first through the Manchu emperors, according to Sharad K. Soni, "who made the specific distinction between 'Outer Mongolia' (外蒙古) and 'Inner Mongolia' (內蒙古)."[95]

The origin of the Mongolian people has not been fully determined, which complicates the development of the ethno-genesis. Yet, as contemporary research acknowledges the identification of early proto-Mongolian tribes can only be made by looking at the various nomadic people who were living in the same "biosphere" as Mongolians are today.[96] However, some of the most commonly cited proto-Mongolians among scholars are the Xianbei, the Xiongnu, the Mengqu Shiwei and the Khitan. The Xianbei (鮮卑) people, for example, existed early as 400 B.C. and were later defeated by the Xiongnu (匈奴).[97] The difficulty is that the nomadic tribes were fragmented and the name "Hu" was applied to several ethnic groups. As historian Nicola di Cosmo concluded, "We can thus reasonably say that, by the end of the fourth century B.C., the term 'Hu' applied to various ethnic groups (tribes, groups of tribes, and even states) speaking different languages and generally found living scattered across a wide territory."[98] But this designation had been lost by the Warring States period (战国时代 475–221 B.C.). The Mengwu Shiwei (蒙无室韦), living in the Northern territories of today's Inner Mongolia and the Khitan of the Liao Empire have been labeled as being the "Menggu" in Chinese-language sources, but it must be taken into consideration

---

94 Jiang, "The Ordos Plateau", 42.
95 Soni, Sharad K. "Looking Back to History: Inner Mongolia under Qing Rule." *Bimonthly Journal of Mongolian and Tibetan Current Situation* 蒙藏現況雙月報第十六卷第五期 1, no. 6 (2007): 32–53. Accessed March 18, 2014. http://www.mtac.gov.tw/mtacbook/upload/09609/0701/2.pdf.
96 Weiers, Michael. *Geschichte der Mongolen* Bd. 586 [i.e. 603]. Stuttgart: W. Kohlhammer, 2004:18.
97 Ibid.15–18.
98 Di Cosmo 2002: Ancient China and its enemies. The rise of nomadic power in East Asian history, p. 130.

that the history of the Liao was only first recorded after 1344, during the Yuan Dynasty.[99]

Nevertheless Hong and other scholars assume that the whole area of Inner Mongolia remained nomadic until the first in-migration of Chinese people during the Warring states period (457–221 B.C.) took place, when first attempts were made to cultivate the region.[100] Following the first unification of China under the Qin Dynasty (秦朝 221–207 B.C.) the central government increased efforts to expand and extend their frontiers into the marginalized peripheries in the North.[101] The Han Dynasty (汉朝 206 B.C. – 24. A.D.) furthered the consolidation of this expansion policy and strengthened cultivation in the region. However, at this time the Xiongnu, after unifying several nomadic tribes and gaining more power in the steppe zones in the North, were a legitimate threat to the Chinese dynasty, which in attempts tried to expand and strengthen their empire into the region.[102] During the later Han, the Xiongnu tribes were split and northern nomads remained nomadic, but stayed within the frontiers of the realm.[103] The strategy utilized by the Han Emperors followed the method of "Controlling the Barbarians with the help of the Barbarians" (以夷制夷).[104] This political strategy was implemented to control the steppe peoples without military force. The ideal of dealing with the "barbarians" without armed intervention wasn't successful and thus further defensive precautions became an alternative.

The following centuries until 589 A.D., at the end of the Chen reign were marked by the diversion of the Chinese empire into different dynasties challenging the sovereignty of foreign peoples.[105] A consequence of this development was that under the pressure of these formations at the North border, the tendency to support a central statehood was strengthened on Chinese side. This impulse to strengthen was not only directed against the human factor, but was also affected

---

99   Weiers, „Geschichte der Mongolen", 18.
100 Jiang, "The Ordos Plateau", 42.
101 Schmidt-Glintzer, Helwig. *Das alte China: Von den Anfängen bis zum 19. Jahrhundert.* 5. Aufl., Orig.-Ausg. Beck'sche Reihe C.-H.-Beck-Wissen 2015. München: Beck, 2008:31.
102 Jiang, "The Ordos Plateau", 47.
103 Schmidt-Glintzer, Helwig. *Das neue China: Von den Opiumkriegen bis heute.* 5., überarb. Aufl., Originalausg. Beck'sche Reihe Wissen 2126. München: Beck, 2009:47.
104 Gao, James Z. *Historical dictionary of modern China (1800–1949).* Historical dictionaries of ancient civilizations and historical eras no. 25. Lanham, Md: Scarecrow Press, 2009:415.
105 Schmidt-Glintzer, "Das neue China", 55.

by physical-geographical factors including the cold winters and the lack of water resources thus requiring the dyke construction at the Yellow River. Between the eastern Han and the Tang dynasty (唐朝 25–581 A.D.) cultivation of the land and the population both declined, and the various Mongolian tribes pushed back their Chinese conquerors, thus allowing the cultivated areas to return to grassland.[106] During the long dynastic history of China the rulers in Inner Mongolia shifted from Mongol to Han and back again, and according to Hong, one of the most noticeable impacts of these shifts was on the environment.[107] In particular the political struggle of this region and the policy of expansion and retreat showed first environmental stress in the form of sandification.[108] Especially the policy that within the first five years of the land reclamation it was exempted from taxes, the Han Chinese settlers practiced a kind of "shifting cultivation", moving from place to place left an exploited landscape behind. First signs of siltation and shifting sand dunes are already described in the *Xin Tang Shu* (新唐书).[109] Only during the Tang Dynasty (618–907 A.D.) it seemed (*Jiu Tang Shu/Old Tang Book* 旧唐书) that there was a time of social freedom, where Mongols and Chinese lived side by side, practicing their different ways of subsistence: grazing and farming.[110]

But with the interlude of Empress Wu (武則天 625 A.D. - 705 A.D.) during the Tang Dynasty the peaceful coexistence had to move and military expansion deep into the Northern territories replaced the hitherto policy.[111] Human activities can be brought in direct relation with the population in the area of the Loess Plateau, and with the north-south border shift between peasants and nomads.[112] When the Mongols established their Yuan state (大元帝国 1279 A.D. - 1368 A.D.), the various Mongolian tribes and the Han-Chinese settlers and merchants coexisted for trading purposes and accelerated the grazing economy.[113] Thus the increase in pastoral nomadism during the Yuan Dynasty led to the agro-pastoral transition

---

106 Ibid.:52
107 Jiang, "The Ordos Plateau", 43.
108 Ibid.:43
109 Ibid.:43
110 Krieger 2009: Geschichte Asiens: Eine Einführung, p. 194.
111 Ibid.:194
112 Xu, Jiongxin. "Growth of the Yellow River Delta over the past 800 Years, as Influenced by Human Activities." *Geografiska Annaler. Series A, Physical Geography* Vol. 85, no. 1 (2003): 21–30:22.
113 Jiang, "The Ordos Plateau", 44.

zone[114] of Inner Mongolia, which continued to retreat further south.[115] In this context, large farmland became grass country, steppes and forests regenerated and served as pastureland, a development which then contributed to the preservation of the surface soil. During these periods of pastoralism soil erosion decreased, as well as the fluvial sediments in the Yellow River.[116]

With the collapse of the foreign rule of the Mongolian Yuan Han-Chinese recaptured the throne. The Ming Dynasty (明朝 1368–1644) was able to hold together the realm, which was unified under the Yuan emperors. Considering that China since the first unification in 221 B.C. until Song dynasty in 1279 was politically fragmented more than half of the whole time, the maintenance of the realm is an important achievement of the Ming rulers. The former military control by feudal landlords was smashed by bringing the military under civil control.[117] The Ming Dynasty became an epoch of autocratic governance.[118] The policies regarding the Mongolian population pursued by Ming China could be described as subversive: not offensively subjecting the Mongols, while simultaneously attempting to destroy their social structures and splitting their tribes.[119] The strategy of "divide and rule", which played into the hands of the Ming, and helped them obtain control over the semi-nomads in southern Mongolia, by allying with them against the "Outer Mongolian" tribes.[120] This ongoing diversion of the Mongolian tribes through internal conflicts resulted in the conquests of the Mongolian homeland by the Manchus and the Russians, who began gaining geo-political influence in this region of Inner Asia.[121] With this defeat, the remaining Outer Mongols under Toghun Temür (1332–1368) retreated back into the steppe.[122] Temür's son, Ayuschiridara, fought the Ming troops until his death in 1378, allowing his people to return to their homelands around its capitol,

---

114 A transition zone is an area, in which diverse properties, either weather conditions, landscapes, ethnics or languages and cultures, are changing slightly or radical from one composition to the other. Inner Mongolia is characterized by the geological transition of High plateaus and the Yellow Riverbed, desert-grassland borders as well as the transition zone of farming-grazing and Han-Mongolian culture.
115 Xu 2003:24.
116 Ibid.:24
117 Schmidt-Glintzer, „Das alte China", 119.
118 Ibid.:119
119 Soni "Looking Back to History", 35.
120 Ibid.:35
121 Ibid.:36
122 Weiers, „Geschichte der Mongolen", 151.

Karakorum.[123] With the end of the Yuan dynasty 1368 the tribe of the Oirats, consisting of four Mongolian ethnic groups (Dzungar (Choros or Olots), Torghut, Dörbet, and Khoshut) gained the mastery over the steppe.[124] In the following two hundred years, the control over the steppe varied from tribe to tribe.[125] During that period, under the reign of Altan Khan (阿爾坦汗 1507–1582) the relation between the Dalai Lama intensified and the Buddhism was propagated.[126] Altan Khan was also the founder of Hohhot, formerly called Köke Khota, the "Blue City". In the 16th century the Jurchen chieftain Nurhaci, descent of the Aisin Gioro clan of the later Qing emperors, reorganized and united various Jurchen tribes and announced himself as the Khan, who also consolidated the Eight Banners[127] military system, administrative units,[128] which were later imposed on the Mongolian tribes in order to unify their organizational structure with the Manchus. The Manchu leader Nurhaci (努爾哈赤 1559–1626) established an alliance with the tribes of the Khalka Mongols to fight against Ming China, emphasizing and constructing a common identity between Mongols and Manchus. This common identity of the two groups became and remained a cornerstone of the later Qing Empire.[129]. In the 17th century the last Mongolian Chahar Khan Ligdan Khutugtu (林丹汗 1592–1634) vainly tried to unify the Mongolian tribes and has undertaken several military offenses in the Ordos and Hohhot area.[130] But the problem was already that the tribes were split into Southern (Inner) Mongolians and Northern (Outer) Mongolians. After the Manchus have won a decisive battle against the Chahar Khan Ligdan in 1634 at the area of Ordos, Inner Mongolia

---

123   Ibid.:155
124   Kollmar-Paulenz, Karénina. *Die Mongolen: Von Dschingis Khan bis heute*. Orig.-Ausg 2730 C. H. Beck Wissen. München: C.H. Beck, 2011:77.
125   Ibid.:78
126   Ibid.:80
127   The Eight Banners functioned as armies, Nurhaci organized his people in household units, initially related to the Jurchen custom. For further information on the "Eight Banners" see: Crossley, Pamela Kyle 1997 "*The Manchus*", p. 6ff.; Crossley, Pamela Kyle 2006 "*Empire at the Margins*", p. 27–52.
128   For further information on the "Eight Banners" see:
      Crossley, Pamela K. *The Manchus*. Cambridge, Mass: Blackwell Publishers, 1997:6
      Crossley, Pamela K., Helen F. Siu, and Donald S. Sutton, eds. *Empire at the margins: Culture, ethnicity, and frontier in early modern China* 28. Berkeley: University of California Press, 2006: 27–52.
129   Soni "Looking Back to History", 38.
130   Kollmar-Paulenz, „Die Mongolen", 85–86.

became part of the Qing state, which in 1644 conquered the Beijing throne.[131] The Chahar tribes of Inner Mongolia attempted to counterbalance the Qing conquerors, while the Khalka Mongols (Outer Mongols) remained neutral, and sent a tribute to the Qing court, which implied submission.[132] The Chahar royal family kept close relations to the Manchu imperial family, intermarriage between prince and princess was a strategy to tie the two royal families together. A family feud between different Khalka groups under Galdan Khan (噶尔丹 1644-1696) and the Dzungar people provoked a war and caused the rural exodus of the Khalka into Qing territory in 1680. But tribal wars between "Inner" and "Outer" Mongols continued until 1688. Following a request of the migrated Khalka to the Manchu ruler Kangxi for military aid in the defense against Galdan Khan, the Khalka accepted the Manchus as their Emperors in 1691.[133]

The geo-political situation of Inner Mongolia during the Qing dynasty (清朝 1644–1911) and the socio-political development of this region, i.e. gaining social welfare and experiencing an expansion of population, helped lead to the transition zone between agriculture and cattle breeding to move back into the North. The integration of Inner Mongolia into China Proper began in the Qing Dynasty (清朝 1644–1911), after the Manchus had conquered neighboring territories of the Western Mongols. During the reigns of Kangxi (康熙 1654–1722), Yongzheng (雍正 1678–1735) and Qianlong (乾隆 1711–1799), the Qing Empire continued an era of stability and prosperity, which strengthened the Inner Mongolian regions after centuries of rival Mongolian tribes. Especially the strong alliance with the Tibetan Lamas supported the Qing's strategy of rule-legitimation outside mainland China.[134] The sixth visit of the Panchen Lama, Lozeng Penden Yeshé, to the Qianlong Emperors summer palace in Chengde in 1780 exemplifies the important role of Tibetan Buddhism during Qing Dynasty.[135] During the reigns of Kangxi, Yongzheng and Qianlong the well-being of the people supported the increase of the population. The rising population, however, also increased pressure on food supplies. The Qing government faced

---

131 Soni "Looking Back to History", 40.
132 Rossabi, Morris. "Bagaryn Shirendyb, et al., History of the Mongolian People's Republic. trans. by William A. Brown and Urgunge Onon. Cambridge Mass: Harvard University Press, 1976. xv + 897 pp. $15.00." *African and Asian Studies* 15, no. 1 (1980): 176. doi:10.1163/156852180X00266:171.
133 Soni "Looking Back to History", 42.
134 Elverskog, Johan. *Our great Qing: The Mongols, Buddhism and the state in late imperial China*. Paperback ed. Honolulu, Hawaii: Univ. of Hawai'i Press, 2008.
135 Ibid.:1,2.

severe distribution problems in many regions, especially in Shanxi, Shaanxi, Hebei and Gansu, which have borders with Inner Mongolia.[136] Natural catastrophes like drought, flood and famine forced the people to leave these provinces and settle in better places searching for food and better new living conditions.[137]

In the following reign, under the government period of Emperor Qianlong (弘曆 1736–1795) large pasturelands and even mountain-slopes were cultivated into fields to meet the requirements of the Han-Chinese migrants. The Great Wall became the demarcation line between pastorals and peasants. The central government wanted to keep Mongols and Chinese separated, in order to prevent Han-Mongolian alliances.[138] The two Mongolian territories Inner and "Outer Mongolia" were ruled by the Qing until 1911, though Inner Mongolia was more directly administered by the Qing court, Outer Mongolia's status had a greater degree of autonomy and was ruled by Mongolian nobles, hereditary Ghengis khanates. The *de facto* authority was given to the lineage of the Tibetan *Jebtsundamba Khutuktu*.[139] To prevent a Han-Mongolian alliance against the Manchu government, the Qing at first sealed off the Mongol dominated regions and forbade mass immigration by Han-Chinese and allowed the Mongolians to keep their traditions.[140] But with the growing population, the Qing adapted their rules and regulations for land reclamation outside the cultivated boundaries.[141] Still keeping in mind that a Han-Mongolian invasion could be a threat to the usurper, the Qing court composed a decree which made Han-Mongolian intermarriage illegal for those trading or settling beyond the Great Wall.[142] During the first decade of Qianlong's reign the Qing government forced the Chinese migration because of the population growth, while also fully aware that the increase of land occupation by Han inflicted upon the changed to the opposite, restricting

---

136 Du, Jingyuan, and Woodworth Max D. "Irrigation Society in China's Northern Frontier, 1860's-1920's.": 7. Accessed June 12, 2013. http://cross-currents.berkeley.edu.
137 For further information on the Great Famines of China, see:
 Yang, Jisheng, and Hans P. Hoffmann. *Grabstein: Die große chinesische Hungerkatastrophe 1958–1962 = Mùbēi*. Frankfurt am Main: Fischer, 2012. http://www.gbv.de/dms/faz-rez/FD1201210063635021.pdf.
138 Jiang, "The Ordos Plateau", 44.
139 Fletcher, Joseph 1978: "*The heyday of the Ch'ing order in Mongolia, Sinkiang and Tibet*". In: John K. Fairbank (Ed.): The Cambridge History of China. Cambridge: Cambridge University Press, p. 351–408.
140 Du and Woodworth 2011: "Irrigation Society in China's Northern frontier 1860–1920", 8.
141 Ibid.:8
142 Ibid.:8

further settlement into the Mongolian heartland.[143] The official decree read, "It is henceforth forbidden to settle and cultivate land in Harqin, Tumed, Aohan, Ongniud banners and in the eight banners of Chahar."[144] These prefectures were under administration of Qing court and situated in the area of today's Inner Mongolia. The Qing government restricted the Mongolians from crossing banner borders, but at the same time they prevent "Outer Mongolia" from population pressure from China Proper.[145] The government also requisitioned former land reclamations.[146] Under the reigns of Qianlong, Jiaqing (嘉慶 1760–1820) and Daoguang (道光 1782–1850) this restrictive policy was upheld. The policy of migration was distinguished by back and forth movement of land reclamation in Mongolian territory. With the beginning of the twentieth century in 1902, the Qing abandoned their restrictive migration policies and opened up the Inner Mongolian grasslands, which were already Sinicized by Han-Chinese merchants and settlers. The influence of Qing governance became less important, as the settlers had made illegal land-contracts with the Mongolian nobles.[147] Nevertheless the Qing court was supporting the settlers with the initiation of irrigation projects, as was done in Hetao.[148]

The contract between "Outer Mongolia" and the Qing Empire became finally invalid after the breakup of the Qing Dynasty. Following the collapse of the Qing Empire, the Republic of China (1911–1949) was established. But the political integrity was lost to independent warlords controlling the various regions, owing the political, economic and military power.[149] The migration to Mongolian territory

---

143 Ibid.:8
144 Ibid.:8
145 Ibid.:8
146 Du and Woodworth 2011: "The court also demanded the requisition of reclaimed land. All agricultural settlers who claimed and cultivated land in Mongolian areas outside the Great Wall in violation of standing law must return to their original place of residence; all land under cultivation must be abandoned or returned to their original owners; failure to abide by this regulation may result in public pillorying, caning, and forcible relocation" (大清會典事例 Da qing huidian shili 1899), p. 8.
147 Weiers, „Geschichte der Mongolen", 223.
148 Ibid.:9
     Thomas E. Ewing. "Russia, China, and the Origins of the Mongolian People's Republic, 1911–1921: A Reappraisal." *The Slavonic and East European Review* 58, no. 3 (1980): 399–421. Accessed July 16, 2015:400.
149 Weiers, "Geschichte der Mongolen", 223.

of Han-Chinese farmers intensified during the Chinese Republic.[150] Outer Mongolia immediately declared itself independent in 1911 with the enthronement of the eighth *Khalkha Jebtsundamba Khutuktu*, also known as the *Bogd Khan*, spiritual head of the Tibetan Buddhism in Mongolia.[151] In 1915 the "Treaty of Kyakhta" between China, (Outer) Mongolia and Russia governed "Outer Mongolia's" autonomy, but as a part of China.[152] But it only remained autonomous until 1919, after General Xu Shuzheng (徐樹錚 1880–1925[153]) reclaimed "Outer Mongolia" as property of the Republic of China.[154] In 1920 as a result of the Russian Civil War, Russian troops supporting the Mongols were defeating the Chinese troops. This event caused the close relation between (Outer) Mongolia and the Soviet Union for the next seven decades. "Outer Mongolia" became sovereign in October 1920 as a revolutionary state under the reign of terror of the Baltic-German General *Roman Nikolai Maximilian von Ungern-Sternberg* (1885–1921)[155]. But this interim control lasted only until March 1921, when a Russian and Mongolian partisan detachment defeated Ungern-Sternberg in Kyakhta.[156] With the death of the eighth Jebtsundampa Khutuktu, Outer Mongolia finally declared its independence and the establishment of the *Mongolian People's Republic* in 1924 and remained until 1992. The collapse of the Soviet Union led to a peaceful democratic revolution and transformed the country into today's sovereign state of Mongolia.[157]

---

150 Williams, Dee Mack. *Beyond great walls: Environment, identity, and development on the Chinese grasslands of Inner Mongolia.* Stanford, Calif: Stanford University Press, 2002:28.
151 Ewing "Russia, China and the Origins", 400.
    Weiers, "Geschichte der Mongolen", 225–228.
152 Ewing "Russia, China and the Origins", 400.
153 Xu Shuzheng was a military warlord in Republican China.
154 Weiers, „Geschichte der Mongolen", 226.
155 Ibid.:229
    Ewing "Russia, China and the Origins", 414.
156 Weiers, „Geschichte der Mongolen", 229.
    Ewing 1980:414–417.
157 Weiers, „Geschichte der Mongolen", 237.

Map 2: Russia, China and the Origins of the People's Republic of Mongolia

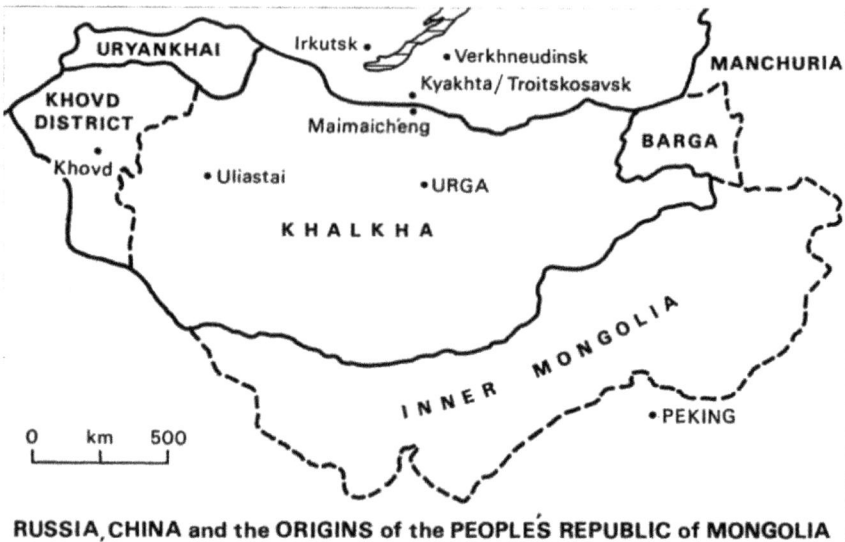

Source: Thomas E. Ewing 1980: *"Russia, China, and the Origins of the Mongolian People's Republic, 1911-1921: A Reappraisal"*

With the took over of the Chinese Communists and their approach to establish the People's Republic of China, Inner Mongolia and Manchuria were controlled with Soviet support. The Communists established Inner Mongolia Autonomous Region (IMAR) in 1947.[158] During that time most of its six million people were nomads, and few depended on agricultural pursuits.[159] But the expansion of cultivation during Mao's *Great Leap Forward* (大跃进) and the *Cultural Revolution* (無產階級文化大革命) led to an acceleration of degraded farmland, grassland, forest and wetland.[160] Mao's call for maximum local self-sufficiency forced the people to plow up farmland, which was not suitable for dryland agriculture. It was the policy of a failed agriculture and deforestation leaving eroded landscapes behind, which were susceptible for storms and increasing wind

---

158   Ibid.:234
159   John W. Longworth, John W. Longworth, and Gregory J. Williamson. *China's pastoral region: Sheep and wool, minority nationalities, rangeland degradation and sustainable development*. Wallingford: CAB Internat. [u.a.], 1993:77.
160   Williams, "Beyond Great Walls", 30.

erosion.¹⁶¹ As Williams 2002 points out, during 1953 and 1979 approximately 21% of the total Inner Mongolian rangeland had to make way for agricultural production.¹⁶²

Before policy measurements attempted to stabilize the situation in the 1980s, the 1960s and 1970s were times when this trend of migration increased rapidly. With the beginning of the reform era in 1979, the population ratio was approximately eleven Han to one Mongol.¹⁶³ A major impact of the national policy occurring with the Han settlement in Inner Mongolia was the forced transfer of pastureland into farmland, even if it was naturally more suitable for raising livestock than for agricultural purpose.¹⁶⁴

While the human population increased, the available rangeland for pastoral use shrunk significantly. New croplands had to be opened up to expand the grain production.¹⁶⁵ This resulted in desertification and soil erosion and an increase of sandstorms.¹⁶⁶ Judith Shapiro (2001) provides a graphic example of the grassland destruction during the Great Leap:

> In Inner Mongolia's Yikeshaomeng [伊克昭盟 *Yikezhao Meng*, today Ordos city], 18 million *mu* of grasslands became desert because of land-reclamation efforts, and sandstorms advanced south, so that every year agricultural land had to be re-seeded, and more than 1 million *mu* of farmland were destroyed.¹⁶⁷

With the post-1978 era came major structural changes in the pastoral regions of Inner Mongolia. The "Property Rights Reforms", so called "double track" or *Double Household Production Responsibility System* (DHPRS) replaced the former communes system, which was based on household teams and brigades, organizing and pooling labor and income. Although this system allowed the state to make huge extraction of the maximum countryside's surplus, the system was abandoned during the Great Famine (1959–1962), because of its inefficiency.

The new system of the *Household Responsibility System* (HRS) gave greater fiscal responsibility for county governments. The introduction of free markets for wool and cashmere had a significant impact on the situation of Inner Mongolia's

---

161   Ibid.:30
162   Ibid.:30
163   Ibid.:28
164   Ibid.:29
165   Shapiro "Mao's War Against Nature", 76,83.
166   Ibid.:107
167   Ibid.:107

livestock numbers, wool output, pastureland degradation and household income, which will be discussed in chapter 4.3.[168]

The forced sedentarization[169] of the Mongolian nomads by the cultivating Han was more exploitative of the rangelands than the pure nomadism practiced before. Before 1985, the nomads practiced short-distance moving within the policy of "collective movement" within the communistic commune system. Thus, herding was allowed for the entire community. After twenty-five years, the Chinese government intervened in the seasonal moving of the Mongolian nomads and transformed it into a semi-nomadic lifestyle.[170] This is because the government considered nomadism as "backward," thus they encouraged the Mongolian herdsmen to settle down in villages and to build brick houses. The herdsmen then remained sedentary from March to August, and moved from October to December to the autumn pastures. When the snow cover became too thick for animals to graze, they moved to the winter pastures.[171] The combination of Han-Chinese settlers migrating to Inner Mongolia and forced settlement strategies for herders led to a decline in nomadism. By 1989, 75 percent of twenty-one million inhabitants were peasants and only 9 percent remained pastoral.[172] In 1990 traditional Mongolian nomadism was already distinct in Inner Mongolia, due to communistic policies and forced settlement. This development had a huge impact on the Inner Mongolian grasslands and in retrospective was responsible for the decline of fertile pastures.

In 1993, Ou Li, Rong Ma and James R. Simpson, interviewed Mongolian herdsmen in a case study of a village in Inner Mongolia on the changes in the nomadic pattern and its impacts on grassland vegetation. Most of the interviewed considered the sedentary life an improvement on the living and working conditions of the pastoralists. However, the herdsmen were already aware of the negative

---

168 John W. Longworth, John W. Longworth, and Gregory J. Williamson. *China's pastoral region: Sheep and wool, minority nationalities, rangeland degradation and sustainable development*. Wallingford: CAB Internat. [u.a.], 1993:47.
169 Sedentism or sedentarization is a term in anthropology to describe the transition from nomadic life to permanent settlement, sometimes forced by the cultivating society "sedentarizing" the nomads to adopt the fixed habitat.
170 Ou Li, Rong Ma, and James R. Simpson. "Changes in the nomadic pattern and its impact on the Inner Mongolian steppe grasslands ecosystem." *Nomadic Peoples*, no. 33 (1993): 63–72. Accessed July 1, 2013:67.
171 Ibid.:68
172 Ibid.:77

impacts on the grassland ecosystem.[173] In terms of environmental awareness, this study confirms that the people, who were affected by these governmental policies reflected early upon advantages and disadvantages of economic development. The herders blamed high animal-production and dense overgrazing for the grassland deterioration. Because of limitations in moving with the herds, the animals grazed on the same pastures over a longer period of time, causing severe grassland deterioration. Even though, the indicators of grassland degradation were identified, the knowledge about the reasons of environmental deterioration didn't led to pro-grassland strategies until the end of the Twentieth Century, when the Grain for Green[174] program started.

The history of Inner Mongolia has shown that various political systems interfered with the traditional nomadic lifestyle. Sedentarization, limited livestock mobility as well as overgrazing through increased stocking rates are factors, which have influenced grassland degradation. In Inner Mongolia herders have been polarized in those with advanced technologies and large herds, able to move over huge distances, benefiting and competing from market conditions and those forced to practice short-term strategies to maintain subsistence.[175] In general one can say that the pastoral economy in China suffered from its peripheral position within in the national economy and from the fact that millions of smallholders were not in the position to influence the market conditions of China's growing economy.

3.2 Inner Mongolia

Inner Mongolia lies in the North of China and shares an international border with Russia and Mongolia. With Manchuria in the east and Xinjiang in the west it represents the Central Asian frontier area of China. Established by Mao's communists in 1947 as an autonomous region (内蒙古自治区) it was constructed on the grounds of the former provinces of Suiyuan, Chahar, Rehe, Liaobei and Xing'an and some Northern parts of Gansu and Ningxia.[176] Inner Mongolia was the first autonomous region of China. Today it represents the third largest province of the

---

173　Ou Li; Rong Ma; James R. Simpson "Changes in the nomadic pattern", 63–72.
174　The Grain for Green Program was initiated by the Chinese government to manage the increasing ecological degradation, reduce the rural poverty and increase the rural household income.
175　Humphrey, Caroline, and David Sneath. *The end of Nomadism? Society, state and the environment in inner Asia*. Knapwell: White Horse Press, 1999:58.
176　Longworth, "*China's Pastoral Region*", 75.

country with an area of 1.2 million square kilometers. (463,000 sq. mi), which is 12 percent of China's total land area. In 1947, approximately 40 percent of the 6 million large population belonged to the Mongolian minority.[177] But today its population is less than 2 percent of total China, approximately 24 million, with a majority of nearly 80 percent Han, 18 percent Mongols and 2 percent Manchu.[178] The official languages are Chinese and Mongolian.

The Inner Mongolia Autonomous Region is often described as one of the "geographically remote, underdeveloped, economically poor and technically underestimated regions."[179] Although the situation has improved since the second half of the Twentieth Century, until 1990 Inner Mongolia was ranked among China's poorest regions.[180] Especially the natural resources of coal and rare earth deposits has pushed Inner Mongolia's per capita GDP with a two-digit growth rate since 2000.[181]

The Inner Mongolian grasslands are the third largest of the world and the area is one of the two major wool growing provinces of China, containing the largest population of sheep, goat and cattle.[182] In the last decades, Inner Mongolia has produced more wool and cashmere than any other province of China.[183] In the 1990s, Inner Mongolia was composed of nearly 87 percent of natural grasslands, of which are 39.2% are already degraded[184]. This huge area supported nearly 37 million animals.[185] Eleven percent of the region's gross output was covered by livestock products. Although the grasslands are huge, the last decades have shown a decline in grassland quantity and quality due to climate change and mismanagement of the natural grasslands, which seems to have accelerated by the

---

177 Ibid.:75
178 Williams, "Beyond Great Walls", 28.
179 Ibid.:1
180 Peng, Mike W. *Global strategy*. 3rd ed. Mason, Ohio: South-Western, 2014:392.
181 "In 2006, GDP reached RMB 479.1 billion, 18.7% higher than in 2005. Figure 12.5.1 shows the GDP breakdown by sector. The industrial sector contributes nearly 50% of the total GDP, while the service and agricultural sectors contribute 37.8% and 13.6%." http://ufa2015.com/portal/sco/members/china/regions/vnutrennyaya-mongoliya.
182 Staff, National R. C., ed. *Grasslands and Grassland Sciences in Northern China*. Washington: National Academies Press, 1992:183–198.
183 Longworth, "China's Pastoral Region", 76.
184 Mengli Zhao. "Grassland Resource and its Situation in Inner Mongolia, China." *Bull. Facul.Agric.Niigata Univ.*, 58, no. 2 (2006): 129–32; 新潟大学農学部研究報告, 第 58 巻 2 号. Accessed February 13, 2014.
185 National Research Council, "Grasslands and Grassland Sciences", 17.

human impact. Today animal husbandry accounts for 45% of the gross output of the agricultural sector in Inner Mongolia.[186]

A significant landscape of Inner Mongolia, beside the rare spots of intact grassland and steppe, is desert or semi-desert.[187] Inner Mongolia shares similar landscapes and animal resources with Tibet and Xinjiang. Most parts of Inner Mongolia lay more than 1000m above sea level, and today most regions are windswept and sandy.[188] The Mongolian Plateau dominates the eastern part of Inner Mongolia and rises up to 2000m. The plateau was uplifted during the Pleistocene Era, which was characterized by the continuous change of cold and warm periods.[189] During that time the Inner Mongolian landscape was covered with coniferous forests and glaciers.[190] Since then many lakes changed into desert areas, which are the source of the extensive Aeolian loess deposits in the southeast, where the Yellow River receives its high amount of silt. In the central part of Inner Mongolia, where the Yellow River makes its huge loop, lies the Hetao Area and the Ordos Plateau. The Ordos is also covered by thick Aeolian deposits, alkali lakes and deserts. In the west of Inner Mongolia is the Alashan Plateau, which also rises up to 1000m above sea level.

Inner Mongolia is divided into several vegetation zones, from the wet and rainy east to the dry and desertificated west.[191] The most profitable and productive grasslands are found in the east, near the border to Manchuria. The fertile agricultural zones are Hetao Irrigation District (河套灌区) and the Ordos (鄂尔多斯), where the fertile soils are nourished by the groundwater flows of the Yellow River basin and higher precipitation per year.

The "typical grassland" [*dianxing caoyuan* 典型草原] of Inner Mongolia can be found in the west of the mountains. It is covered with "a mid-height, heavy-cover community, [which] runs across the vast castanozem[192] plateau west of

---

186 HKTDC Research (Ed.) 2015: "*Inner Mongolia: Market Profile*": http://china-trade-research.hktdc.com/business-news/article/Fast-Facts/Inner-Mongolia-Market-Profile/ff/en/1/1X000000/1X07T7RO.htm.
187 Ibid.:77
188 Longworth, "China's Pastoral Region", 77.
189 National Research Council, "Grasslands and Grassland Sciences", 21.
190 Schlütz, Frank. *Palynologische Untersuchungen über die holozäne Vegetations-, Klima- und Siedlungsgeschichte in Hochasien (Nanga Parbat, Karakorum, Nianbaoyeze, Lhasa) und das Pleistozän in China (Qinling-Gebirge, Gaxun Nur): mit 7 Tabellen / Frank Schlütz*. Berlin: Cramer in der Gebr.-Borntraeger-Verl.-Buchh, 1999.
191 Ibid.:17
192 Castanozem refers to one of the 30 soil groups classified by the Food and Agriculture Organization (FAO). Available at:

the mountains and is dominated by *Stipa grandis, Aneurolepidium chinense,* and *Agropyron michnoi.*"[193] The diversion of grassland follows the regional landscape conditions and includes the "dry grassland" [*ganhan caoyuan* 干旱草原] further west beyond the Yingshan mountains and the "desert grassland" [*huangmo* 荒漠] on the Ordos plateau.[194] "Dry grassland" is marked by short grass and "desert grassland" by a scarce vegetation of shrubs and grass patches between the brown soils.[195] Mongols call the eight Chinese gravel desert zones *Gobi* (戈壁), whereas the four sandy deserts are called by the Chinese name of *shamo* (沙漠). The gravel deserts are distributed in the west of the steppe and characterized by moving dunes.[196] Due to the strong winds, the materials from arid Northern China are transported from north-west through the grassland to the south-east, from gravel to sandy. China's "sand belt" stretches 5,000km from west of Taklimakan Desert to the east of Gobi. The expanding desertification of the dynamic deserts affect 400 million people, living in this area. The grasslands of Inner Mongolia are extensive, but both the quantity and the quality of this resource are in danger of decline.[197]

Clearly, the growing economic development of this region comes along with the degradation of the natural ecology. Inner Mongolia's environmental damage is caused primarily by anthropogenic pressure. Urbanization processes and industrialization have further caused a decline in fertile soils all over China. The demographic shifts in population growth since 1950 caused an increase of food demand. Simultaneously agrarian land is shrinking. Intensive cropping, as a result of the attempts to secure an increasing food demand, is further reason for the decline in high quality farmland and grassland. This development has also contributed to the deterioration of Inner Mongolian landscapes. The vicious circle of deterioration has also been exacerbated further due to the grazing herds, which destroy the remaining pastureland because of their exceeded livestock numbers. Increase of farmland, marginalization of fertile pastures and the process of land fragmentation have further worsened the environmental situation. Severe water shortages have also been caused by low rainfall and few natural water deposits. Additionally, water-waste together with inefficient water-use has become a catastrophe for farmers and herders, as they dread low yields. The ecologic result

---

http://www.fao.org/docrep/W8594E/w8594e05.htm#chapter%202:%20key%20to%20the%20reference%20soil%20groups.
193 Ibid.:17
194 Ibid.:18
195 Ibid.:18
196 Williams, "Beyond Great Walls", 24.
197 National Research Council, "Grasslands and Grassland Sciences", 18.

of overgrazing, poor agriculture, chronic water shortage are the reason that both grassland and farmland are endangered by severe droughts and desertification.

3.2.1 Dryland Insight

A comparative look at examples of different sites of Inner Mongolian landscapes will help to explain and understand the variety of land-use and its accompanying environmental degradation. The basis material for insight into the diverse ecological sites of Inner Mongolia came during field research from September to December 2013, during which 300 people from all over Inner Mongolia were interviewed. Recognizing that different places face diverse ecological problems confirmed that it is essential to present an overview on Inner Mongolian landscapes for a full picture of the environmental crisis facing the region today.

Degradation of pasturelands as well as farmland is a major issue in Inner Mongolia. The native grasslands of Inner Mongolia are famous throughout China, especially the Ordos grasslands, Keerquin (科尔沁) and Chifeng (赤峰) Prefecture grasslands. These regions are still major livestock raising districts, but during the last decades these areas have slowly transformed into deserts, and the grassland has been largely destroyed.[198] Beyond these well-known districts, far more pastureland and farmland is being seriously degraded through soil erosion. In 2005 Zhao described that large erosion occurs in "Chifeng City, Xing'an League, Hohhot City, Baotou City, Ordos City, with erosion accounting for more than 70% of their total area".[199] Li, Wenjun, Professor of Environmental Management at the College of Environmental Science and Engineering at Beijing University has declared that half of Inner Mongolia's fertile grasslands have deteriorated and this frightening trend seems to only continue, with dire consequences for Inner Mongolia's economy, ecology and people.[200] The herders are affected by grassland productivity loss, due to grassland decline, which directly influences the local economy, as the meat and dairy production still accounts for approximately 10% of Inner Mongolia's GDP.[201]

---

198 Longworth, "China's Pastoral Region", 82.
199 Zhao "Grassland Resource and its situation", 131.
200 Wenjun Li, Lynn Huntsinger. "China's Grassland Contract Policy and its Impacts on Herder Ability to Benefit in Inner Mongolia: Tragic Feedbacks." *Ecology and Society* 16, no. 2 (2011): 1–14. Accessed June 8, 2015.
201 HKTDC Research (Ed.) 2015: "*Inner Mongolia: Market Profile*": http://china-trade-research.hktdc.com/business-news/article/Fast-Facts/Inner-Mongolia-Market-Profile/ff/en/1/1X000000/1X07T7RO.htm.

In the Chifeng City prefecture lies the Wengniute (翁牛特旗) Banner.[202] Once it was an area of substantial forests and rich grasslands supporting a modest population. Now it is part of the desert-steppe in the Keerquin Sandy land.[203] By 1988, 87 percent of the pasture land in Chifeng City had been destroyed, and is today the worst affected area of desertificated grassland in Inner Mongolia.[204] The climate of this region is semi-arid with average precipitation ranging 300–500mm, mainly in the summer months of June, July and August. Sand dunes occupy 90 percent of the land and fertile pastureland was distributed in the 1960s to collective authority for hay production during the winter.[205] Historically this region has been home to a dense migration of Han farmers, which began in the nineteenth century, setting up a pattern of ever-increasing pressure on the natural environment.[206] This trend, continued during the Twentieth century, with a sudden increase in the 1950s and 1960s, when Han-farmers were settled to the border areas to cultivate grain. This has led to a reduction of the traditional Mongol herders' grazing areas, thus leading to destruction of the local environment.[207]

Wulanaodu Village, approximately 100km north of Chifeng City, an area which receiving received Mao's recognition for its model pastoral commune in 1958, faces severe siltation problems today.[208] Currently, the village is fighting against salinization and alkalization of the land. Sand storms and moving sand dunes are spreading, and vegetation is being destroyed, showing obvious signs of an endangered environment. Land fragmentation and barbed wire plots of pastures are widespread. Wulanaodu Village has already been part of research done by Dee Mack Williams in 1996, where the policy of grassland enclosures and its impact were discussed, in the context of environment, grassland degradation, ethnic identity and with broader focus on local development and grassland policies.[209]

250 km north-west of Wengniute is the Dianxing Caoyuan[210] of Xilin Gol (锡林郭勒盟) League. Xilin Gol has a greater variety of soils and a higher altitude. Therefore the temperature is lower, but it has the same annual precipitation. Both

---

202 Williams, "Beyond Great Walls", 18.
203 Ibid.:21
204 Longworth, "China's Pastoral Region", 82.
205 Williams, "Beyond Great Walls", 25.
206 National Research Council, "Grasslands and Grassland Sciences",18.
207 Ibid.:18
208 Williams, "Beyond Great Walls", 23.
209 Williams, Dee Mack. "Grassland enclosures." *Human Organization* 55, no. 3 (1996): 307–14. Accessed October 14, 2014.
210 Typical grassland, described in chapter 3, "midheight and heavy-covered with grass".

regions were influenced by the migration of Han, the increasing livestock and the expanding farmland. Nevertheless, Xilin Gol has less density of population and is not as suitable for agriculture as Wengniute Banner. Xilin Gol thoroughly shows signs of degradation, but most of the "typical grassland" is still intact.[211]

Siziwang (四子王旗) Banner, also served as the research site for the evaluation of potato farmers, offers another different view of the grassland situation of Inner Mongolia. One hundred and sixty kilometers north of Hohhot (呼和浩特) it is situated in the northern part of Yinshan Mountain on the Inner Mongolian Plateau. It covers an area of 25,500 sq.km and has a population of 209,000. Most of its inhabitants depend upon agricultural activities like farming and animal husbandry.[212] The transition zone of agriculture and animal husbandry is just North of Hohhot. The weather is best characterized as continental, hot with low rainfall in summers and dry and cold during winters. The winter period lasts from October to April and the summer is short from June to September. The annual precipitation is relatively low with 100–300mm.[213] Climate and rainfall rise and fall across the natural border of the Daqing Mountains and along the northern and southern slopes rainfall is sufficient and the soil is fertile enough to plant wheat, potatoes and oats. Nevertheless from south to north, the natural environment becomes more raw and dry to sustain agriculture and therefore is suitable for pastoralism.[214] Wind erosion in Siziwang is very high, as Han Guodong 2011 explains, "Strong winds and poor vegetation cover result in significant amounts of wind erosion every spring from the fine unconsolidated surface soils typical of the region."[215] The Siziwang grassland productivity suffers from a "summer soil-moisture deficit and limited soil nutrients", due to the climate. But additional pressure through overgrazing occurred under the HRS, when individual households were settled and allocated to specific tracts of land.[216] Siziwang

---

211 National Research Council, *"Grasslands and grassland sciences in northern China"*, 19.
212 Han, G. D., Li, N., Zhao, M. L., Zhang, M., Wang, Z. W., Li, Z. G., Bai, W. J. P., Jones, R., Kemp, D., Takahashi, T., and Michalk, D. "Changing livestock numbers and farm management to improve the livelihood of farmers and rehabilitate grasslands in desert steppe: a case study in Siziwang Banner, Inner Mongolia Autonomous Region." *Development of Sustainable Livestock Systems on Grasslands in North-western China. ACIAR Proceedings*, 2011, 80–96.
213 Ibid.:81
214 Ibid.:81
215 Ibid.:81
216 Ibid.:82

was also affected by the governmental policies, which started with the *Great Leap Forward* (1958–1960) and the following two decades, when the farmers were forced to intensify cultivation of farmland into the mountains.[217] This caused the marginalization of the Mongols into the Northern periphery or their shift from animal husbandry to agriculture. If the years were warm and moist this policy succeeded, but in drier and colder seasons the crops failed.[218] Population density and stocking rates have increased the pressure on grasslands since 50 years. Siziwang grassland has been classified as a zone of fragile ecology. Political attempts to expand farmland further into the north has increased environmental problems in this area. Currently, regional policies still try to limit the agricultural expansion by clear allocation of pastures and cultivated land.[219] In the further north regions of Siziwang, pastoralism is recommended and supported by governmental policies. But in the agricultural region, closer to Hohhot, mixed economies can also be found. Today, Siziwang is famous for its potato production and is a major producer of potatoes in Inner Mongolia. To adjust the uncertain and insecure living conditions, farmers from this region also raise animals and let their sheep feed on the reduced grasslands. Another way to secure additional income is to rent yurts to the urban tourists during holiday seasons.

Often the families' economic background does however determine the family business. For example, families, which practiced nomadism through generations, tend to pass this kind of subsistence along to their children. But today in the context of long-term immigration of farmers a mixed economy of agro pastoralism has developed. Often interethnic marriages in Inner Mongolia have further contributed to the ethnic mosaic of Han-Mongolian coexistence, also evident in the agricultural sector, where Mongolian farmers and Han-Chinese pastoralists characterize the scenery. Both Han and Mongols are working in agriculture and animal husbandry. There is no clear separation between ethnic descent and subsistence economy. It seems that people choose their subsistence income sovereignly, and not necessarily based upon any historic-ethnic connection. How does this affect Inner Mongolia's environment? The assumption is that a farmers' impact on environment is higher than the nomads', who is in close communion with nature and follows the seasonal cycle.[220] In general the nomadic culture traditionally doesn't exploit natural grasslands and therefore

---

217 National Research Council, "Grasslands and Grassland Sciences", 20.
218 Ibid.:20
219 Han, G. D. et al., "Changing Livestock Numbers", 82.
220 Zhang, MunkhDalai A., Elles Borjigin, and Huiping Zhang. "Mongolian nomadic culture and ecological culture: On the ecological reconstruction in the agro-pastoral

prevents desertification; whereas, agrarian culture extracts natural soil deposits and facilitates desertification.[221] But looking at the history and latest development of Inner Mongolian grasslands since 1960, with the rise of intensive agriculture and food demand, some Chinese grassland scholars believe that both agriculture and animal husbandry contributed to the degeneration of the Inner Mongolian grasslands.[222]

3.2.2 Hetao Plain and Hetao Irrigation District

In contrast to the classical grassland structures described previously, the Hetao area demonstrates a different ecology in that way that the area alongside the Yellow River Loop is nourished by the water of the river and a large artificial canal-system. Constructed in the early nineteenth century, this area is referred to as Hetao Irrigation District (HID). Therefore a specification of Hetao Plain and "Hetao Irrigation District (河套灌区)" is essential. The Hetao Plain is surrounded by mountains, the Alashan in the west, the Lüliangshan in the east, the Yinshan in the north and the Great Wall in the south.[223] It is part of Ningxia, Inner Mongolia and Shaanxi and today it is mainly divided in two areas: *Xitao* (西套) and *Dongtao* (东套). The name "Hetao" (河套) originates from the course of the Yellow River and the huge loop it makes in the north.[224] Hetao's history is marked by the transition of the Mongol and Han society and its overlapping land-use systems. The demarcation line between the two traditional subsidy economies changed with the demand of fertile agriculture or pastures, and was controlled by the respective government until 1949. This shift-induced interference led to different land-use practices and different local production systems. The geographical location is remarkable for the history of intercultural contact between two economies, specifically, the sedentary agriculture of peasants and non-sedentary pastoralism of nomads.

---

      mosaic zone in Northern China." *Ecological Economics* 62, no. 1 (2007): 19–26. doi:10.1016/j.ecolecon.2006.11.005.
221 Ibid.:19
222 National Research Council, "Grasslands and grassland sciences in northern China", XIV.
223 Du and Woodworth 2011: "Irrigation Society in China's Northern frontier 1860–1920", 1.
224 王天顺 Wang, Tianshun. 黄河文明 河套史: *Huang He wen ming He tao shi*. Di 1 ban. Beijing: Ren min chu ban she, 2006.

During the Ming and Qing dynasties, Hetao was separated into three parts: Houtao (後套), Qiantao (前套) and Xitao (西套).[225] The following map shows the whole area of Hetao Plain (河套平原), including the location of HID and the cities of Hohhot, Baotou, Linhe and Dengkou.

Map 3: *Larger Area of Hetao (including Houtao, Qiantao and Xitao) during Ming and Qing Dynasty*

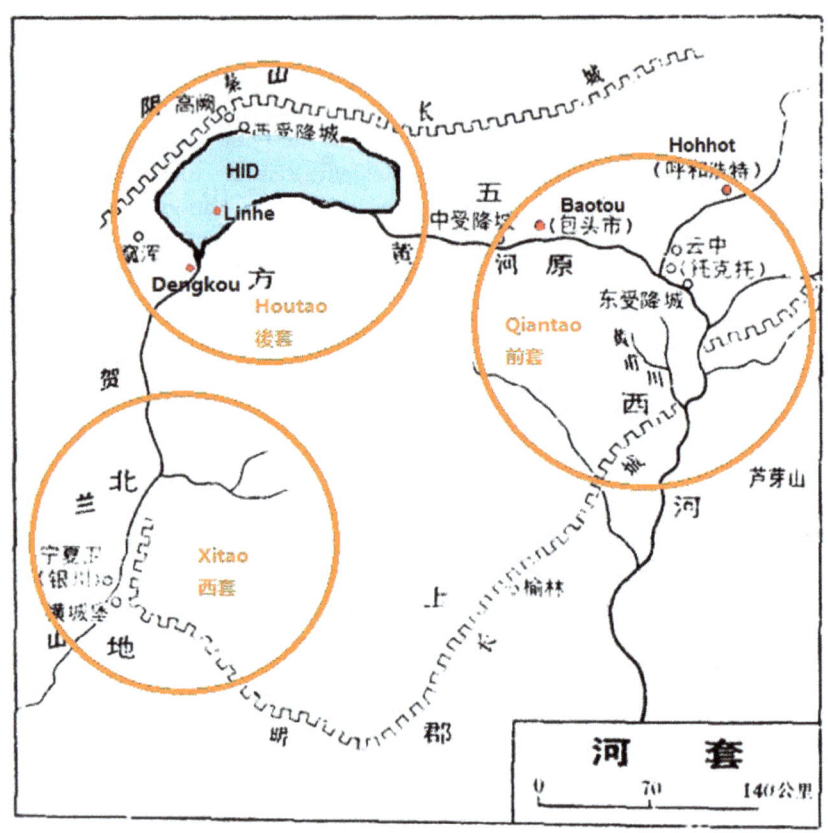

Source: Cilia Neumann (2016), *Historical Map of Hetao*, modified map

The second map of Hetao Plain illustrates the *status quo ante* (Late-Qing agricultural areas) of the Hetao area during the Ming and Qing dynasties. Here the

---

225 Ibid.3.

62

distribution of nomadic areas (nom.), as well as agricultural settlements (agr.) is marked. Obviously agricultural areas were still small, despite the fact that the settlement policy was supported by the central government at the time. Additionally, the map shows a very small area of remaining forest, between the Ordos (鄂尔多斯右翼后旗) and Wulateqi (乌喇特旗). However, most of the area, as can be seen from the illustrated map, remained nomadic and has been color-coded as "nomadic places".

Map 4: Agricultural Areas in Hetao during Ming and Qing Dynasty

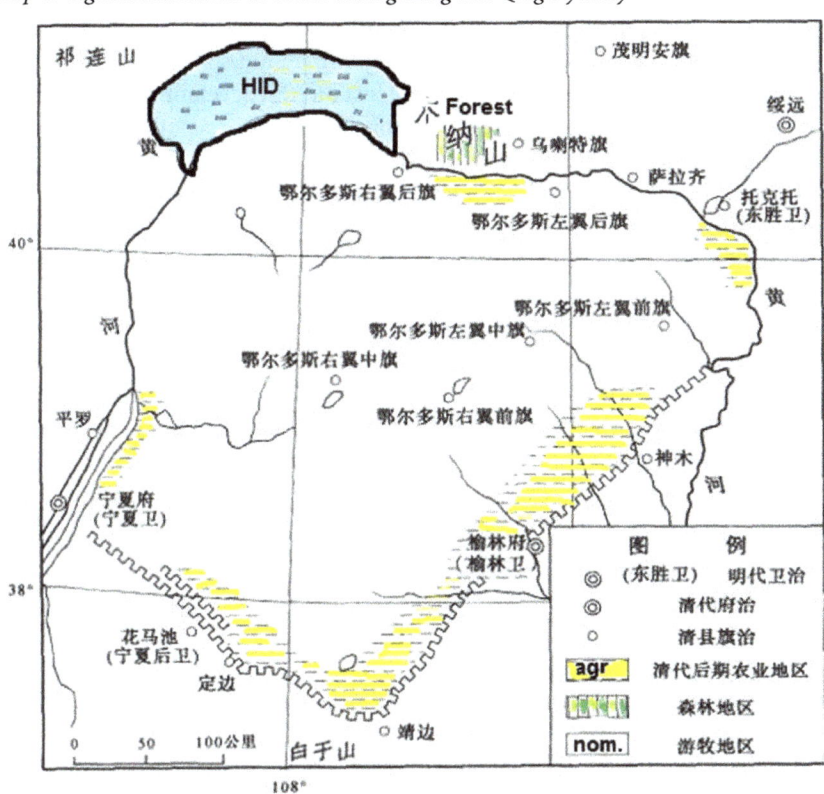

图 5-6 明清时期鄂尔多斯高原和河套平原农林牧分布图

Source: Cilia Neumann (2016), *Historical Map of Hetao*, color-coded and modified map[226]

---

226 http://bbs.tiexue.net/post2_3990109_1.html.

With the increase of Han merchants settling in the Hetao area, land reclamation also accelerated, and Hetao's economy finally made the transition from pastoralism to sedentary agriculture.[227] Han migrants moving into the Hetao area settled in current-day Ulan Chabu (乌兰察布盟), Hohhot (呼和浩特), Baotou (包头), Ordos (鄂尔多斯) and parts of Bayan Nur (巴彦淖尔盟) and Wuhai (乌海). They were colloquially known as the settlers who "exited the west gate" (走西口).[228] Change came, however, under the reign of Kangxi (康熙) emperor, during Qing government from 1661–1722, as he began a push for the reclaiming of wasteland, which also led to the simultaneous development of water resource exploitation.[229] In the beginning of the Kangxi Era, most of the new migrants lived in the area on a seasonal basis, arriving in the mild spring and leaving for the winter.[230] Only with permanent settlements did the development of the HID start. Referring to the work of Karl Wittfogel and his thesis of the "hydraulic civilization"[231], the development of Hetao's canal system suggests that the non-state actors (merchants, Catholic Church) played a decisive role in the construction of this irrigation system from 1860–1920. It further suggests that they were additionally responsible for the agricultural expansion in this frontier area. The early canal construction in Hetao has benefited from the historical Yellow River course shift in 1850, when the river diverted into the great Inner Mongolian loop.[232] The full-scale irrigation project was started 30 years later by Wang Tung-chun, a Han-merchant who settled to Hetao.[233] In 1881 he constructed the first of ten canals (Yihe canal) flushing water from the Yellow River into the fields, which enabled the successful long-term agriculture in Hetao.[234] A report from 1919 by the Hetao area canal administration noted that

> The Church seized opportunities to penetrate Houtao, occupy land, and demand compensation in land from the Mongol banners for any incident sparked by their missionary

---

227 Du and Woodworth 2011: "Irrigation Society in China's Northern frontier 1860–1920", 3.
228 Ibid.:7
229 Ibid.:8
230 Ibid.:7
231 Wittfogel, Karl A. *Oriental despotism: A comparative study of total power*. 1. ed. Vintage Books V-701. New York NY u.a.: Vintage Books, 1981.
232 Ibid.:11
233 Dabringhaus, Sabine. *Territorialer Nationalismus in China: Historisch-geographisches Denken 1900–1949* Bd. 2. Köln: Böhlau, 2006:194.
234 Ibid.:194

work. The Church has thus accumulated land parcels of over 2,000 *qing*[235] in the western Houtao and may, in fact, hold over 10,000 *qing*.[236]

The Church actively contributed to the irrigation system in Hetao with financial assistance and by the late Qing period, the Church had also organized the digging of several irrigation canals in Dengkou, area of Bayan Nur.[237]

To visualize an overview of the ancient canal-system constructed by Wang Tungchun, the following map provides detailed insight into the canal system of HID, which is still in use today. The Yellow River course as well as the main canals is colored in.

Map 5: Hetao Irrigation District with its Canal System

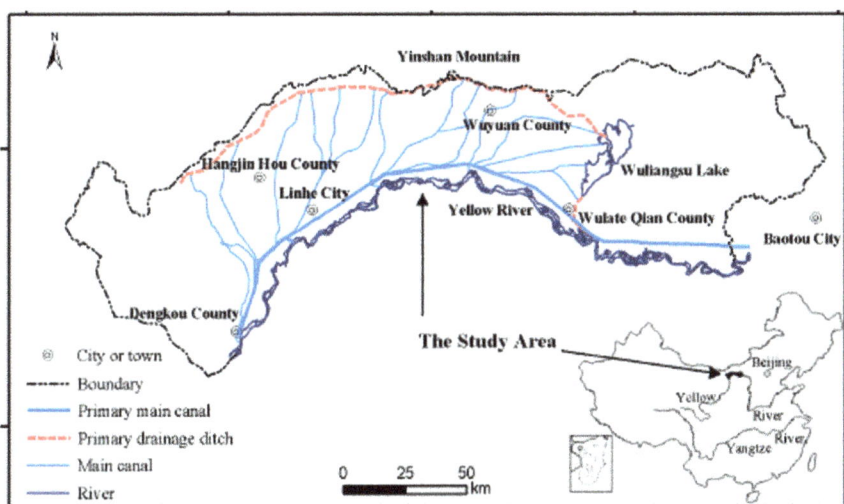

Source: Zhang, Y., Y. Luo, W. Zhao, and Hatina M. "Land use dynamics and landscape change pattern in Hetao irrigation district." *The International Society for Optical Engineering* 7841 (2010): 2

The agricultural history of Hetao has only intensified over the last century. However, the decrease of cultivated farmland might be attributed to the land-use

---

235   Ancient measurement for an area unit equivalent to 100 mu.
236   Suiyuan Gazetteer Archive Editorial Board 2007.
237   Du and Woodworth 2011: "Irrigation Society in China's Northern frontier 1860–1920", 14.

change since 1986, which continued until 2002.[238] Only in conjunction with the "Grain for Green" policy, which has been transforming slope-farmland into forests, primal farmland has been reconverted into grassland in order to prevent further erosion and desertification.[239] Many farmers seem to support this policy because the government has been providing financial support and granting subsidies for farmers who stop tilling their land since 2002.[240] Other evidence which suggests this interpretation is the increase in soil salinization and poor land quality, which has led farmers to abandon their farmland.[241]

Since the reform era of Deng Xiaoping (邓小平) starting in 1987, investments in science and technology have been made to increase the production and yields.[242] Mainly the institutional changes coming along with the implementation of the HRS were attributable to the progress of Chinese agricultural productivity.[243] With the full implementation in 1984, contributions of institutional changes are no longer relevant on productivity growth. Technological development, such as new technology and seed varieties, market information, and access to education are factors which benefitted the farmers to improve their technical efficiency in production.[244] But during the early 1990s, water conservancy programs combined with the lack of knowledge stopped the local agricultural development and caused severe damage to the ecosystem. Especially the forced cultivation of oil crops in combination with pesticides and fertilizers increased salinization in the Hetao Plain.[245] The 1980's have already showed early signs of environmental distress causing the stagnation of further economic development of Hetao. At the same time the growing desertification and degradation became a major constraint. Eighty percent of the area was affected by salinization, which caused a decline in the fertility of the soils from a one quarter to one

---

238 Zhang, Y., Y. Luo, W. Zhao, and Hatina M. "Land use dynamics and landscape change pattern in Hetao irrigation district." *The International Society for Optical Engineering* 7841 (2010): 1–9:6.
239 Ibid.:6
240 Ibid.:6
241 Ibid.:6
242 王关区 Wang, Guan Q. "河套草原农业的发展方向和模式: Hetao Agricultural development, direction and investment pattern." *经济社会*, 1995, 48–49.
243 Mao, Weining, and Won W. Koo. "Productivity growth, technological progress, and efficiency change in Chinese agriculture after rural economic reforms: A DEA approach." *China Economic Review* 8, no. 2 (1997): 157–74. doi:10.1016/S1043-951X(97)90004-3.
244 Ibid.:169
245 Wang, "Hetao Agricultural Development", 48.

half. Forested areas decreased more than 8 percent, due to extreme deforestation, which in turn increased the damages of natural wind erosion and resulting in an increase of desertification.[246] This resulted in a weak economy and a shortage in the agricultural production.

Until the mid-1990s, the Hetao Plain has been in a transmission phase from traditional to modern agriculture and from a semi-commodity to a market economy.[247] The agricultural resource utilization, labor productivity, processing proliferation rate were much lower than in developed countries The labor productivity rate was equivalent to about one percent of the US.[248] Thus the development of a rural market, the circulation of agricultural products and their expansion in HID were quite slow. The economic reasons behind this were a low net-income for farmers, a lack of long-term investments, and no increase in sales markets. There was urgent need to support the agricultural development in Hetao through technological innovations, research and development (R&D) strategies, integrating the farmers into the local markets. The first response to this economic recession was that the Chinese government started a program in 1998 in an attempt to solve the chronic water-shortage in HID, with the aim to stop the increase of salinization.[249] With industrial progress and the urbanization of Inner Mongolia, the degree of Yellow River pollution and the pollution of Hetao has increased as well. Especially the water pollution, in particular, has increased massively.[250] Since the 1960s and 1970's the amount on arsenic pollution in the Hetao Plain, due to mining has caused severe health problems for the local population.[251] Today the issue of soil nitrate (N) leaching off the irrigated

---

246　Ibid.:48
247　河套平原的农业处于传统型向现代型, 半自然半商品经济向本场经济的过渡阶段.属于石前化特征明显, 生态经济效益较低下, 科技水平相对格后的农业, 是生态经济系统不良的传胍有机农业和现代石油农业的 "杂交型农业". Ibid:48.
248　河套平原农业的资源利用率, 劳动生产率, 加工增殖率, 商品销售率等较发达国家低待多, 劳动生产率仅相当于美国的百分之一左右. Ibid.:48
249　Yang, Yuting, Songhao Shang, and Lei Jiang. "Remote sensing temporal and spatial patterns of evapotranspiration and the responses to water management in a large irrigation district of North China." *Agricultural and Forest Meteorology* 164 (2012): 112–22. doi:10.1016/j.agrformet.2012.05.011.
250　郭宇. 确保黄河绿水张流, 河套绿色驻.对巴彦淖尔市环境保护工作的思考. 内蒙古工作2005 年第7期. Guo Yu. "Ensuring the Yellow Green Flows. Green water in the Loop – Environmental protection in Bayonner" 2005: 27.
251　Zhang, Dongsheng Ma, Xiongxi Hu, Hui. "Arsenic pollution in groundwater from Hetao Area, China." *Environmental Geology* 41, no. 6 (2002): 638–43. Accessed July 28, 2015. doi:10.1007/s002540100442.

farmlands due to application of N fertilizers and large volume irrigation has increased the amount of nitrate in the groundwater.[252]

Today the HID covers 2 million square kilometers of which are 80 percent are irrigated.[253] Currently, Hetao is a large-scale irrigation area, flushing 294,000,000 cubic meters of water annually from the Yellow River for the irrigation of 500,000 ha of cultivated land. This artificial irrigation system made Hetao famous for its crop, vegetable and oil production and is considered with the attitude of being "Beyond the Great Wall Granary".[254] Forty percent of Inner Mongolia's agricultural production, specifically grain, oil, and sugar, takes place in the fertile fields of Hetao Irrigation District.[255] HID has become one of the most important grain bases in China, as well as being one of the most relevant economic sectors of North China.[256] The continental weather conditions, without high temperature falls and nearly frost-free periods, also makes the planting of vegetables possible. However, these natural climatic conditions result in far less precipitation than required, and therefore irrigation is essential for cultivating vegetables and crops.[257] Whereas the rest of Inner Mongolia is fighting against drought and unproductive soils by cultivating drought-resistant crops like potatoes, the manmade canal system of the HID has given this area the character of an oasis. Still, without water from Yellow River, agriculture could not have been possible and the climate and man-induced changes have had a significant impact, reducing

---

    Guo, Xiao j., Yoshihisa Fujino, Satoshi Kaneko, Kegong Wu, Yajuan Xia, and Takesumi Yoshimura. "Arsenic contamination of groundwater and prevalence of arsenical dermatosis in the Hetao plain area, Inner Mongolia, China." In *Molecular Mechanisms of Metal Toxicity and Carcinogenesis*. Edited by Xianglin Shi et al., 137–40. Boston, MA: Springer US, 2001.

252 Feng, Zhao-Zhong, Xiao-Ke Wang, and Zong-Wei Feng. "Soil N and salinity leaching after the autumn irrigation and its impact on groundwater in Hetao Irrigation District, China." *Agricultural Water Management* 71, no. 2 (2005): 131–43. doi:10.1016/j.agwat.2004.07.001.

253 Wang, Lunping, Yaxin Chen, and Guofang Zeng. *Nei Menggu he tao guan qu guan gai pai shui yu yan jian hua fang zhi: Irrigation drainage and salinization control in Neimenggu hetao irrigation area*. Di 1 ban. Beijing: Shui li dian li chu ban she, 1993:48.

254 Brief Introduction to Inner Mongolia Autonomous Province. Available at: http://news.xinhuanet.com/english/2003-04/02/content_815391.htm. Accessed February 2016.

255 Ibid.:48

256 Zhang "Land use dynamics and pattern in Hetao", 2.

257 Yang, "Remote sensing", 112.

fertile soils and increasing desert areas.²⁵⁸ The impact of agricultural activities and the climate change have allowed the runoff of the river decline significantly over the last five decades.²⁵⁹ The harsh environment in the greater Hetao area had limited productivity and caused unstable yields, therefore cultivation has been difficult and was limited to the irrigated areas, which are represented by the HID. The canal construction of HID and its irrigation method supports that the water from the Yellow River washes the salt off and the seeds can germinate. But this also carries the risk of inappropriate water and land-use practices, which are largely responsible for causing secondary salinization, due to false flooding irrigation, inaccurate drainages and extensive cultivation.²⁶⁰ Additionally, the intensive livestock raising and overgrazing has caused the severe degradation of the last remaining intact grasslands in the HID.²⁶¹ Yinhui Zhang, Yi Luo and Wenwu Zhao in "*Land use dynamics and landscape change pattern in Hetao irrigation district*" (2010) describe the decrease of farmland, grassland, desert and marshland, while forested land, water areas, residential areas and mid-density grassland have increased, though apparently at the cost of dense grassland.²⁶²

The Yellow River is the lifeline of the HID, and essential for prosperous agricultural development. Therefore it is necessary to improve the allocation and utilization of water resources to strengthen the water conservancy.²⁶³ Several different research studies regarding the current environmental problems have been done, especially in regard to water resource management, soil salinity control, and crop response to water salinity. The rehabilitation of the irrigation district and the protection of the entire environment are necessary. Thus Hetao requires the implementation of a Yellow River pollution control and emission control system.

With the growing upper middle class in China, a new sector has captured China's market: organic food. This development is a chance for Inner Mongolian farmers to connect with the new economic conditions which allow for a new sort of sustainable production methods. The HID and its fertile plains currently symbolize the "Green Movement" of Inner Mongolia and are becoming a

---

258 Zhang, Y., Y. Luo, W. Zhao, and Hatina M., "Land use dynamics and landscape change pattern in Hetao irrigation district", 6.
259 Yang, "Remote sensing", 112.
260 Ibid.:7
261 Ibid.:7
262 Zhang, Y.; Luo, Y.; Zhao, W.; Hatina M., "Land use dynamics and landscape change pattern in Hetao irrigation district", 1.
263 Ibid.:27

national, pollution-free agricultural production base.[264] With the economic integration into globalized markets and the transition to organically produced food, including a sustainable production chain, the Hetao area has the ability to double its value with the brand of being a "green ecological agricultural region."[265] By respecting the laws of nature, optimizing economic structures, adjusting the industry, changing the mode of economic growth and improving resource utilization and the recycling of waste, Guo Yu (2005) sees the first steps to develop an eco-economy in Hetao.[266]

---

264 郭宇. 确保黄河绿水张流, 河套绿色驻.对巴彦淖尔市环境保护工作的思考. 内蒙古工作2005 年第7期. Guo Yu. "Ensuring the Yellow Green Flows. Green water in the Loop – Environmental protection in Bayonner" 2005: 27.
265 Ibid.:27
266 Ibid.:28

4. Pastoralism versus Agriculture: Different Land-Use Varieties

Questions omnipresent when dealing with environmental degradation in Inner Mongolia include: What causes grassland degradation, soil erosion, and water shortages? And: Is it agriculture or pastoralism or a combination of the two which is responsible for land loss and desertification? The following chapter tries to find answers to these questions. The situation today is far from the romantic perception of vast Inner Mongolian grasslands, which are sprinkled with Mongolian yurts and horse riders dressed like Genghis Khan, who follow their herds for hundreds of kilometers. On the contrary, the Inner Mongolian grasslands had to make way for industrial purposes like mining, urban development, and agricultural use. Pastoralism became part of the domestic economy, which must serve national demands. Inner Mongolia's dairy and meat production relies on the remaining pastures and grassland resources. This chapter will reveal the problems in the discussion of traditional land-use practices associated with sustainability and the requirements of a modern industrialized society, typically denounced for being responsible for Inner Mongolia's grassland destruction. Therefore the two agitators, agriculture and pastoralism, are in focus, including their economic, social, and political development, all of which have affected herders and farmers in Inner Mongolia.

The question as to why Inner Mongolia in the seventeenth century became attractive for migration might be answered by the circumstance that the land was not already settled by a peasant society. The land in this region was perceived to be available for settlement and the fertile soils were waiting to be cultivated. The Han-Chinese motives for settling in Inner Mongolia were to escape from famine and poor living conditions, with the assumption that their quality of life would improve settling in new areas. Though the potential for ethnic conflicts due to the unfair distribution of resources and land-use were great, they were largely avoided due to a clear distinction between and mutual understanding of both Mongolian nomadism and Han-Chinese agriculture.[267] During the Qing reign the Han and Mongols thrived side by side. The space at hand was large, and peasants and herdsmen both lived from the same sources of grassland, water, and soil. But the growing population and immigration of Han Chinese forced

---

267 Williams, "Beyond Great Walls", 70.

the Mongols to move with their herds further north to the remote, unsettled, and unfertile pastures of Inner Mongolia. As such, ethnic conflicts both historically and today are often related to these migration-related issues.[268] The influence of Han-migration into the nomadic pastoral areas especially in Inner Mongolia is of high relevance and has had a fundamental impact on nomadic culture and land-use.[269] These changes to Inner Mongolian nomadic culture were accompanied by the long-time existence of Han-Chinese immigrants, who followed a sedentary lifestyle, which is contrary to nomadic culture.[270]

Typically Chinese scholars, scientists, and officials are of the opinion that the cause of Inner Mongolia's degradation and desertification can be linked to past and present anthropogenic forces.[271] In particular, the severe degradation of Inner Mongolia can be attributed to anthropogenic pressure.[272] Though climate changes and physical processes did have an initial impact on the general environmental deterioration of the Inner Mongolian grasslands, humans and their way of changing the landscape for agricultural purposes has contributed to the enlargement of the degradation.[273] The extension of deserts in Inner Mongolia is also widely attributed to human activity.[274] Lattimore has written about the impact of the Han-colonization in Inner Mongolia, explaining, "The type of colonization created by the rapid building of railways demanded quantity rather than quality," which was done to supply themselves with goods, and the colonists started to cultivate the grassland.[275] Despite low precipitation and heavy climatic

---

268 Xing, Li, and Qi Xing. *Nei Menggu qu yu you mu wen hua de bian qian: The Nomadic Cuture Change in Inner Mongolia*. Di 1 ban. Beijing, 2013:185.
目前世界上几乎所有国家的族群冲突都与历史上曾经发生过或目前在进行的迁移活动有关，**因此迁移成**为研究一个地区族群关系现状及其演变过程的重要切点.

269 Ibid.:184

270 Ibid.18 内蒙古区域游牧文化的变迁是伴随着大规模的长期的汉族移民发生的、这是一个社会现实.我们不仅要研究其发生的形态、而且要研究其发生的原因.农耕的生计方式是定居生活的模式、安土重迁不仅仅构成其生活模式、而且形成其文化性格.

271 Williams, "Beyond Great Walls", 27.

272 Williams, "Beyond Great Walls", 28.

273 Ibid.:27

274 For further information on desertification in China see: "*Drylands development and combating desertification*". Available at: http://www.fao.org/docrep/t4410e/t4410e00.htm.

275 Lattimore, Owen. *The desert road to Turkestan*. Kodansha globe. New York, N.Y: Kodansha International, 1995.

conditions the exponential population growth marked the beginning of the natural misbalance,[276] and according to the Committee on Scholarly Communication for the People's Republic of China 1992 report, "too many people and too many animals are pressing too hard on a fragile ecosystem."[277] Thus, pastoralists became a minority and their rangelands decreased continuously.[278] From 1930 to 1950 the animal population diminished, which was largely attributable to the disastrous campaign of "Wipe out the Four Pests" [*chu si hai* 除四害] during the Maoist era.[279] Deforestation and pastureland degradation during the *Great Leap Forward* and *Cultural Revolution* took their toll on the environment.

Not only periods of political struggle contributed to the changing of the Inner Mongolian environment but also a clash of cultures. Two dominant cultures had a large influence and are responsible for changing the natural ecology of the area, namely the Han-Chinese and Mongols dominated and shaped the steppe-ecosystem of Inner Mongolia especially since Qing dynasty.

4.1 Culture Clash: Han and Mongols – Distinction and Delineation

The role of culture cannot be underestimated for its impact on human perceptions and behaviours. Religion and beliefs as internal factors influence the mindset and the environmental attitude of a population. The question addressed here, is how this cultural systems can affect humans interactions with their natural environment. Thus the different grades of environmental degradation under different human societies are a visual proof of human impact on the environment.[280] Therefore evaluating the local culture helps to understand a society and its land-use practices, their different ways of interacting with their environment and different attitudes towards nature.

---

276 Ibid.:27
277 Staff, National R. C. *Grasslands and grassland sciences in northern China: A report of the Committee on Scholarly Communication with the People's Republic of China, Office of International Affairs, National Research Council*. Washington: National Academies Press; National Academy Press, 1992:33.
278 Williams, "Beyond Great Walls", 29.
279 Ibid.:82
280 Jiang, "The Ordos Plateau", 41.

Han-Chinese Heritage

Contemporary Chinese are ancestors of the autochthonous population, which settled in North China during the Neolithic age.[281] The name Han-Chinese dates back to Han-Dynasty (206–220 A.D.) and correlated more with the Chinese language and culture, than with the genetic descent.[282] For thousands of years the Han have cultivated their land and certain characteristics are common with all peasant societies[283]:

1. They all show a deep connection with their homeland.
2. Settlement is a major aspect of farming societies all over the world. To succeed in agriculture it is important to build up villages and settle near the cultivated fields. However, this lack of mobility makes far-trading impossible. Therefore agriculture was mainly for subsistence.
3. Because peasant societies live in close proximity to each other, they often rely on social orders and codes of conducts. Rules and regulations which declare the daily living within the society, building up social ethics, and certain hierarchies are essential to maintain the community.

Traditionally the life cycle of peasants is related to the birth-and-death cycle of the people and depends on the seasonal cycle of farming.[284] Han people show a strong loyalty to their land and community. According to Hong, "People grow out of their land and return to it, like leaves returning to their roots."[285] Traditionally Han try to protect their homeland and pass it on to their descendants, therefore it is important to protect and keep the quality of the land intact. One possibility is the shifting cultivation, which means the land is cultivated temporarily, then abandoned and allowed to regenerate, while the group moves to another field. After a period of three to five years the field recovers and is cultivated again. Under normal conditions, shifting cultivation is an efficient method of food production, adapted to the natural habitat. But in the case of an unexpected rapid population growth, regeneration of the fields is not possible because of the fatal circle of slash and burn, expansion of fields and shorter fallowness. With the in-migration of

---

281   Haarmann 2004 "Kleines Lexikon der Völker", 103.
282   Guter, Joseph 2004: Lexikon zur Geschichte Chinas", 178.
283   For further information on the characteristics of different substistence patterns see: Beer, Bettina, ed. *Methoden und Techniken der Feldforschung*. Ethnologische Paperbacks. Berlin: Reimer, 2003:101.
284   Jiang, "The Ordos Plateau", 54.
285   Ibid.:55

Han peasants, long before intensive cultivation in this region began, this method of cultivation together with a population boom also led to erosion and siltation in the area of Ordos and other regions of Inner Mongolia.[286]

Only the permanent settlement made a social life within the group possible and certain specialists were able to develop their skills in religion, handicraft, and politics. Large-scale tillage farming and the establishment of villages laid to the foundation of the later peasant society. The implementation of technological achievements like the harrow for breaking soil and artificial irrigation made this evolution possible. These renewals enabled population growth and increased the areas of settlement. It can thus be assumed that generating a food-surplus is linked with the arising of modern states and more complex societies. As early as during the Proto-Chinese period simple slash and burn farming was practiced, as it was the most successful way of feeding larger groups of a few hundred people. All over the world the advanced cultivation technologies of harrow and plough lead to population increase, production surplus and in the end to the development of organized states.[287] Until 1949 it was still common to use ploughs and hoes, cattle and bare hands to cultivate the land.[288] People were dependent on natural conditions and so different methods of farming according to the environment developed and were recognizable throughout China.

Water irrigation systems as well as rain-fed farming[289] are responsible for the development of intensive farming. But with the development of newer technologies (threshing and sowing machines), intensive farming expanded and developed in industrialized agriculture, which is able to increase yields to the maximum, by using pesticides, fertilizer and high-tech equipment for field work.

An important ideology of peasants, especially in China, is that although the embeddedness into national and international economies has a higher value than before, the older generations still tend to show the most respect to those living from land and doing hard labor.[290] This creates a new dilemma: despite the circumstance that in Han tradition peasant living is considered to be of high moral value and standing, the modern Chinese society does not value this way of living.

---

286 Jiang, "The Ordos Plateau", 48.
287 Gernet, Jacques. *Die chinesische Welt: Die Geschichte Chinas von den Anfängen bis zur Jetztzeit*. 1. Aufl. Suhrkamp-Taschenbuch 1505. Frankfurt am Main: Suhrkamp, 1988:23.
288 For further information on the development and history of Chinas agriculture, see: Perkins, Dwight H. 2013: "*Agricultural Development in China 1368–1968*".
289 Rain-fed farming is agriculture relying on rainfall for water.
290 Jiang, "The Ordos Plateau", 55,56.

Thus contemporary farmers are of low social status within China's population.[291] Therefore farming on a small-scale is socially and economically dishonorable. On the other hand rural people are disadvantaged regarding education in comparison with their city counterparts.[292] Today peasants recognize a higher education level in order to get a skilled job. But entering the university affords the passing of the national examinations, which hinders most farmers. Often the assumption of the older generation is that school is useless and only serves the purpose of social advancement.[293] On the other hand most of the Inner Mongolians interviewed stated that they push their children into schools and universities to give them the opportunity to find a job in the cities.

Mongolian Nomadism

The **Mongols** traditional land-use is raising livestock and grazing on the semi-arid grasslands of North Asia.[294] This type of economy follows the natural moving of the herds searching for water resources and fertile pastures.[295] Therefore high mobility is required, trekking from one plot to another. The ability of free moving with the herds is essential to pastoralism. Predominantly, these societies live a subsistence-lifestyle, using every part from the animals they raise. Nevertheless pastoralists need to maintain economic relations with peasants for intercultural trade. Trading animal products in exchange for grain or technical products is one example for the intercultural economic interaction of pastoral societies with peasants.[296]

---

291 Ibid.:55,56.
292 Ibid.:55,56.
293 Ibid.:57
294 Ibid.:50
295 Detailed information on nomadism is provided by the works of Rolf Herzog (1919–2006), German anthropologist and Fred Scholz (1939), a Germany geograph.
Herzog, Rolf. *Sesshaftwerden von Nomaden: Geschichte, gegenwärtiger Stand e. wirtschaftl. wie sozialen Prozesses u. Möglichkeiten d. sinnvollen techn. Unterstützung / Rolf Herzog.* Köln: Westdt. Verl, 1963.
Scholz, Fred, ed. *Nomaden, mobile Tierhaltung: Zur gegenwärtigen Lage von Nomaden und zu den Problemen und Chancen mobiler Tierhaltung; 20 Beiträge.* Berlin: Das Arab. Buch, 1991.
Scholz, Fred. *Nomadismus: Theorie und Wandel einer sozio-ökologischen Kulturweise.* Erdkundliches Wissen 118. Stuttgart: Steiner, 1995.
296 For further information on nomadism, the author refers to the works of Rolf Herzog and Fred Scholz.

The social organization of pastoral societies is typically a tribal structure organized in clans. Various household-units belong to a certain tribe. Lineage is based on a patrilineal kinship system built around a common ancestor and this lineage is the basis for property rights. Although pastoral societies are marked by sharp labor division – men are mainly responsible for herding the livestock, whereas women get engaged in handicraft and the diary production – the status of the women is not particularly low in these societies. Throughout history, Mongol women have played an important political role, often on behalf of their sons, which were too young to reign.

Mobility has also been historically essential for pastoral groups to organize themselves into confederations, particularly in the case of war.[297] This ability aided Genghis Khan in uniting the various Mongolian tribes between 1184 and 1204, to become the Khan of all Mongols in 1206, thus establishing the Mongol Empire and the basis of the Yuan Dynasty (1279–1368).[298] In the early sixteenth century and before, the Mongols were typically believers of shamanism, but they adopted Buddhism in the late sixteenth century. Their religious philosophy of life harmonizes with nature which can help to explain their contemporary relationship to the environment around them.[299] The lives of people and livestock corresponded with the natural habitat. When pastures were exhausted, they moved to another. The grasslands had sufficient time to recover with the following season.

Compared to Han peasants, Mongolian herdsmen have traditionally spent more money and energy to reinvest and protect nature.[300] They have also given less attention to hierarchy and social status, which can also impact how a society interacts with its natural environment. As a group, it is recognized that they live from the land and pastures, and thus the only way to improve their living is through a sustainable method of natural resource management.[301] A further reason for this contemporary habitus might also be the lower economic opportunities pastoralists have, especially the difficulties faced to find a job outside the grazing-herding sector. Depending on their traditions and method of income, there were fewer chances to work in another field. Furthermore, older generations often did not speak Chinese and had less formal education than the Han.[302] Today this situation might have changed, and Mongols have more opportunities

---

297 Jiang, "The Ordos Plateau", 167.
298 Kollmar-Paulenz, „Die Mongolen", 22.
299 Kollmar-Paulenz, „Die Mongolen", 15.
300 Jiang, "The Ordos Plateau", 50.
301 Ibid.:58
302 Ibid.:60

to work in a wider variety of businesses. With the economic change of the 1980s and the development of large-scale livestock production, the ecological balance has been, and continues to be disturbed. Large-scale livestock production brings new problems: overgrazing, water shortage; *dzud* are just a few. Additional feed and hay-production for animal sustenance increased the grassland yields, but this also led to exploitation. The land's value was transformed from being a cultural symbol for Mongols to a source of economic income for the titular Han nation.[303]

According to all historical accounts, the changes in land-use and the increase of demographic pressure due to Han colonization during the seventeenth and eighteenth century escalated in the ecological destruction of the steppe zone.[304] Chinese settlers cultivated the fertile land despite inadequate irrigation and heavy storms often supported by Mongolian nobles, who maintained land-use contracts with settlers and catholic missionaries. These missionaries have moved to Inner Mongolia to establish monasteries during the eighteenth century to promote and distribute Christianity in Northern China. The monasteries also held close relations with Mongolian Nobles for trading purposes.[305] With the increase of population after 1949 and the accompanying economic policy changes, the traditional method of land-use became impractical. Therefore the government policy of "grain first" during the *Cultural Revolution* essentially forced the people back to traditional farming.[306] Nearly every piece of land had to be cultivated, even if it was more suitable for grazing. One consequence of this was the step-by-step erosion of soil. The fertile soils were then blown away by the strong winds and the sand began to cover the land.[307] Extensive cultivation for subsistence purposes then turned to mass-mobilized intensive agriculture, exploiting nature in all its facets. With the movement of Chinese settlers into the periphery of the area, and the extension of land-cultivation beyond the Great Wall, the biosphere of Mongolian and Chinese culture, together with their traditional way of living, diminished successively.

---

303   Ibid.:54
304   Williams, "Beyond Great Walls", 28.
305   Taveirne, P. *Han-Mongol Encounters and Missionary Endeavors: A History of Scheut in Ordos (Hetao) 1874–1911*. Leuven University Press, 2004.
306   Jiang, "The Ordos Plateau", 48.
307   Williams, "Beyond Great Walls", 29.

4.2 Political Changes in Inner Mongolia

Since the establishment of the Autonomous Region of Inner Mongolia in 1947, and the foundation of the People's Republic of China in 1949, China's political line has been characterized by instability and vicissitude between the political campaigns. To gain the trust of the Mongol population in Inner Mongolia, the Communist Party declared in 1948 that the grasslands would be protected and the Mongol traditions should be respected.[308] With the political take-over of the Party in 1949, the regional government reestablished the grasslands as a protected area in order to secure livestock grazing and the nomadic way of life.[309] Returning the Inner Mongolian grasslands to the Mongols was step towards reducing the inter-ethnic conflict between the Han-Chinese and Mongol population, which has existed since Han settled in the Mongolian territory during the sixteenth and seventeenth century of Qing dynasty.[310]

With the phase of political re-establishment and economic restoration during the years 1949 to 1957, grassland protection was one of the key issues.[311] While in most parts of China people were released from their suppressed feudal position and were allowed to begin reaping the rewards of their own labor, Inner Mongolia implemented a policy known as the "three-No's": "no execution of herd-owners, no redistribution of livestock and no class differentiation among herders."[312] This meant that Inner Mongolia was not integrated into the national politics of the socialist regime, as its regional grassland policies showed considerable sensitivity to regional history and Inner Mongolian culture. While landlords in other parts of China were persecuted and their properties were redistributed, regional regulations in Inner Mongolia also encouraged grassland restoration. Within the first five-year plan from 1953 to 1957, the government's plan was to transform the traditional Mongolian nomadism into a managed, profitable business.[313] During this period Mongolian herders built their first settled houses.[314] Older approved methods were combined with new practices of grassland management: wolf eradication, well digging, and livestock shelters.[315] The result was that herd

---

308   Jiang, "The Ordos Plateau", 643.
309   Ibid.:643
310   Ibid.:644
311   Ibid.:644
312   Ibid.:644
313   Ibid.:644
314   Ibid.:644
315   Ibid.:644

populations increased rapidly.[316] Although this period at least in the beginning brought a better life to the rural population, it also led to an acceleration of environmental destruction, especially in the years following.[317]

By reshaping the landscape, the Maoists believed that man could also be reshaped and reformed. 1958 to 1965 was a period in which economic development was accompanied by political movement, exploiting natural resources for the "development first – treatment later" policy.[318] The parole of "greater, faster, better, and more economically" (多快好省) characterizes how necessary it was for Inner Mongolia to reach the same economic standards as other parts of China. Within this political revolution the slogan "battle with nature", or rather "combat nature", became the standard program. The initial methods for reconstructing and 'combating' the Inner Mongolian grasslands were shrub planting and the transformation of deserts, and a campaign to turn these areas into a green 'oasis' spread all over the region.[319] Where trees had been cut centuries before, the first attempts were made to undo the devastating consequences of the large deforestation,[320] while in other parts of China large-scale tree cutting continued the complete deforestation process of entire tracts of land.

The following decade from 1966 to 1976 was disastrous for both the people and the environment as economic or political problems were ignored, and the country focused on a politicized class struggle.[321] *The Great Leap Forward* and the *Cultural Revolution* were national catastrophes for China's social, economic, and political development. Transforming China into a communist, collectivized society, both campaigns were responsible for land-wide environmental degradation. Not until the reform era between 1977 and 1978, did the Chinese government undergo a political turn towards economic reformation, nor did the government develop a strategy to eliminate social disparity or to support environmental remediation. Only in 1973 the first national environmental protection meeting was held and for the first time China's environmental issues were raised.[322]

---

316 Ibid.:644
317 Ibid.:644
318 Jiang, "The Ordos Plateau", 89.
319 Jiang, "Grassland management", 641–53.
320 Elvin, Mark. *The retreat of the elephants: An environmental history of China.* New Haven: Yale University Press, 2006:19–40.
321 Jiang, "The Ordos Plateau", 89.
322 China Council for International Cooperation on Environment and Development (CCICED): Developing Policies for Soil Environmental Protection in China. Hg. v. CCICED.

As a consequence the country addressed pollution factors and environmental problems, mainly water and air pollution. After regional investigations the State Council established an environmental management system which was observed by the leading Group of Environmental Protection and its office, in charge of environmental protection.[323] Between 1979 and 1992 increasing attention was being paid to environmental pollution. The Environmental Protection Law of the People's Republic of China, implemented in 1979 and the Constitution of the People's Republic of China, issued in 1982, as well as the Land Administration Law of the People's Republic of China, issued in 1986, all incorporated policies for environmental protection. Since 1992, China has developed a Five-Year Environmental Plan, in line with the Five-Year Social and Economic Development Plans, which includes various provisions and subsidies to strengthen environmental protection management.[324] In 1996, the Ninth-Five-Year Plan highlighted the increase of environmental destruction and called for the establishment of environmental management and legislative systems to minimize the land-wide pollution. These efforts later cumulated in the latest "Grain for Green" policy initiated in 1999 and launched nationwide in in 2002.[325]

4.2.1 Inner Mongolia's Variety of Land-Use

The traditional land use in Inner Mongolia has always and still varied with the climate. When the zone is warmer and wetter, the more likely it is that agriculture exists, as well as extensively cultivated farmland. Historically, the expansion of farmland has always been driven by irrigation schemes, the immigration of Han Chinese peasants into the grassland peripheries, and by the political pressure from Beijing.[326] With global changes in weather conditions, the pure

---

323 Ibid.:378
  Further information on Environmental development in China under: http://www.fao.org/docrep/p4150e/p4150e01.htm.
324 Workshop on Environment, Resources and Agricultural Policies in China (2006): Environment, water resources and agricultural policies. Lessons from China and OECD countries; [these proceedings bring together papers from the Workshop on Environment, Resources and Agricultural Policies in China, held in Beijing, on 19–21 June 2006]. Paris: OECD.
325 Liu, Can, and Bin Wu. "Grain for Green Program in China: Policy Making and Implementation?" Briefing Series, China Policy Institute, University of Nottingham, April 2010:2 https://www.nottingham.ac.uk/cpi/documents/briefings/briefing-60-reforestation.pdf.
326 National Research Council, "*Grasslands and Grassland*", 26,27.

agricultural use of land has given way to a mixed pastoral-agricultural economy. Most of the farmers, and in this case not cash crop farmers, also raise animals in order to gain additional income. They do not live in remote grassland areas, but rather closer to the agglomerated areas of cities where resources are more accessible and their own farmland can be cultivated in combination with livestock holding. Pastoralism in the fertile, moist areas is scarce and limited. Only in the northern, cold, and dry areas of Inner Mongolia extensive pastoralism can be found. This economy is characterized by seasonal migration between summer and winter camps where the livestock can be fed and watered. In winter periods, the animals are fed with feed from storage. The pure nomadism, which was once the traditional economy of the Mongolians, no longer exists in Inner Mongolia.

The limitations imposed by the rough climate are reflected in the distribution of people and animals.[327] The population density in northern China is small (15 percent of the total Chinese population), whereas the numbers of grazing livestock is 60 percent of the total national livestock. The economic activity of Inner Mongolia goes hand-in-hand with the distribution of the different ethnic groups living there, which provides further insight into the migration movement of the Han over last centuries. The rough and harsh regions of Xinjiang, Qinghai, and Tibet are less frequently settled by Han-Chinese than the fertile plateaus of Inner Mongolia.[328] Political campaigns supporting the migration of Han were motivated in the belief that by removing the population pressure in the south via transferring people to northern and eastern remote and underdeveloped parts of China would help alleviate the overpopulation problem.

Since 1949, the Mongolian herdsmen have undergone enormous political and economic changes due to the various campaigns implemented by the Chinese government in Beijing. The following sub-chapter will illustrate the difficulties for Mongolian herders to cope with the various political systems. Two large campaigns, namely the *Land Reform* in 1949 and the *Great Leap Forward* in 1958, shifted land rights for agricultural purpose or livestock from private elites (landlords) to peasants or collectives in the Maoist era.[329] Today, a third party has become a part of the equation of land-rights holdings: Often properties are owned by private producers, but still under the control of the state.[330]

---

327 Ibid.:27
328 Ibid.:27
329 Ibid.:8
330 Ibid.:8

4.2.2 Impact for Mongolian Herders

Throughout history nomadism has developed in regions with harsh climates, meaning scarce precipitation, and rough geological structures like mountains and high plateaus. Some examples of these regions include the steppe areas and deserts of Central Asia and Africa. In fact, human adaption to natural habitats is an effective way to survive and conserve the own species. Therefore grassland, pampas, deserts, and steppes are areas where nomadism fits in the ecological niche. The advantage of nomadism in these regions is that nomads have traditionally followed water sources and pastures in a seasonal moving pattern in order to maintain the grassland ecology and to better cope with unpredictable natural phenomenon or natural disasters. Throughout Inner Mongolian history, versions of these patterns of migration have been witnessed, and it is only following the Communist takeover that these patterns have begun to drastically change due to political interference into the nomadic lifestyle via the regulation of moving patterns and forced sedentarization of herders. But until the takeover of the Communists in 1949, the nomadic lifestyle had only undergone minor control measures and remained essentially autonomous in terms of their moving patterns.

The following sections will explain the impacts of political reforms and regulations in the nomadic lifestyle such as the limiting of herders from moving, increasing stocking rates, and their effect on Inner Mongolian grassland ecology. Grassland ecology was more or less intact from the Yuan to Ming Dynasty and only with the governance of Qing in the eighteenth century did the grassland became subject of larger political interest, coming along with migration policies and land-use rights.

The first changes in the nomadic production model occurred under the reign of Genghis Khan and in the following Yuan Dynasty when fixed grassland areas were distributed among different tribes. This policy reduced tribal conflicts and established the basis for the Mongolian realm.[331] The Mongolian legal basis code, *Yassa*, included provisions for the sustainable treatment of the grasslands and prohibited digging and setting fire on the grasslands.[332] It is interesting that the *Yassa* rules and regulations regarding the treatment of the grasslands were passed orally through generations for over 700 years, despite the different policies which came in the following centuries under different dynasties. These different policies attempted to regulate grassland use and often neglected the importance of

---

331 Wu, Zhizhong 2008: "Pastoral Nomad Rights in Inner Mongolia", 15.
332 Ibid.:16

a sustainable ecology.[333] Thus the continuity of the *Yassa*, despite official policies stating otherwise, shows that the nomadic traditions are deeply rooted in the Mongolian culture and hard to eradicate.

The banner system implemented by the Qing Dynasty was one of the first to cause damage to the grassland environment. The first land rights for pastoralists were established under the Qing in 1740, which meant that the right of rangeland was controlled by regional despotism of "animal landlords" and tribal leaders.[334] Specifically, by combining the Manchu and Mongolian banners through intermarriage, the nomads' radius of moving with the herds was reduced to a mere 150 kilometers.[335] This resulted in a reduced moving cycle, which caused the first signs of grassland degradation, because the pastures did not have sufficient time to regenerate. In addition, with the further promotion of Lamaism by the government, monasteries were built and many Mongols became lamas, instead of raising livestock and practicing nomadism.[336] Though during some Qing reigns (Kangxi, Yongzheng and Qianlong) there existed regulations to ban Han-migrants from claiming Mongolian land, the Qing government supported the policy of ethnic dissipation by encouraging the in-migration of 30 million people, twenty times the size of Mongolian population.[337] It is remarkable that most of these Han-migrants turned into herders to practice a mixed subsistence.[338] Despite these major changes in settlement policy, the nomad production model of moving within the ancestral pastures was not affected. But the grasslands shrunk and the population rose. Nomadism was marginalized and farming government-funded, thus the grassland degraded rapidly.[339] The human damage during this period caused the disappearance of wetland, decline in surface waters and the expansion of the Ordos Desert.[340]

During the seventeenth and eighteenth century China had already faced its first social and economic collapse due to several factors including the exploitation of natural resources, a population explosion, as well as exhausted grain storages. This serious situation during the Qing dynasty was the basis for the coming revolution of the communist regime. With the victory of the Chinese

---

333 Ibid.:16
334 Longworth, "China's Pastoral Region", 43.
335 Wu, "Pastoral Nomad Rights in Inner Mongolia", 16.
336 Ibid.:16
337 Ibid.:16
338 Ibid.:16
339 Ibid.:17
340 Ibid.:17

Communist Party, the next policy-induced grassland damage occurred, when the publicly owned rangeland, meaning the land belonging to the people living on the grasslands, was distributed as pastoral property and passed on to villages or to the state for community purposes.[341] This period lasted from 1947 to 1950 and was followed by further reforms in livestock ownership from 1947 to 1957. The Rangeland Ethnic Reform Law in 1947 was implemented in order to establish an equilibrium for all herders gaining free access to the grassland. It was not however determined how the livestock should be treated.[342] The national Chinese Agrarian Land Reform regulated the redistribution of livestock to poorer herders rather than to rich landlords, who were in focus to have enriched themselves by claiming large areas of land.[343] In Inner Mongolia the central policies did not succeed because of experiences in the Chifeng prefecture, where livestock was killed rather than redistributing it to poor households. The provincial government decided not to redistribute the livestock from rich herders in Inner Mongolia but rather chose instead to increase the wages for herders responsible for the livestock of a landlord.[344]

Nevertheless, from 1952 to 1958, Inner Mongolia also underwent a transition into an era of collectivism and cooperatives. The Communist regime was convinced that, well-organized human labor is the key to China's economic development. Being organized in cooperatives helped the herders by pooling labor during peak seasons. Through the system of communes, the rural communities were able to accomplish large water irrigation projects, establish small factories and produce goods, which would increase the general income.[345] In the years after 1958, the cooperatives turned into communes, which indicated that the livestock and grasslands are owned by a village, holding the contracts with the households.[346] Parallel to the communes, the government introduced the *Household Responsibility System (HRS)*, which was the prototype of the today's *Household Production Responsibility System (HPRS)*.[347] Its purpose was to enable poor

---

341   Ibid.:43
342   Ibid.:43
343   Ibid.:43
344   Ibid.:44
345   Detailed information on the Communes system available under: http://afe.easia.columbia.edu/special/china_1950_commune.htm.
346   Ibid.:44
347   Ibid.:45

and remote households to profit from the proceeds of their production and to return property rights from collective to private.[348]

With the disruptive forces of the Cultural Revolution in 1966, pastoral property rights and private ownership were considered to be capitalistic behaviour. The central Chinese government's ideology was to "free" rural China from the despotism of landlords and the bourgeois class. A manifestation of this desire came in the mass-mobilization of the Red Guards and the forced elimination of private property in most parts of China. Most of Inner Mongolia's livestock was released to the open grassland rather than redistributing it to the people in the communes.[349] During 1968 and 1969, more than 40 percent of livestock died due to harsh climate and dzud, so that by the end of the Cultural Revolution only small quantities of livestock could be returned to private ownership.[350] This policy reversal, however was very variable, in Inner Mongolia households were allowed to own livestock, but in Gansu, herders were prohibited from owning livestock until the implementation of the present HPRS in 1983.

With the reform era following 1978, the Communist Party implemented major reforms in rural areas. The "double contract" HPRS involved the return of the livestock to private households, "which in return contracted with the commune."[351] The households were allocated the right for ownership of pasture lands. The herders were thus in charge of holding two contracts with the commune for owning livestock and grassland. In Inner Mongolia this policy meant that herders had to share a so-called "double responsibility", which combined the distribution of livestock and the leasing of land for the feed production. This policy reached its peak in 1996 when the policy of "promoting stock-raising business on the basis of the grasslands" helped to push the local economy.[352] The number of livestock increased rapidly, and reached over 10 million in the Xilin Gol league in Inner Mongolia because the people ignored the grasslands herding capacity as a result of this economy-pushing policy. At the same time, however, it led to serious problems for the herders, namely, income-loss, grassland degradation, and decrease in livestock.[353] In 1996 land conversion, wire fencing

---

348 Wu, Harry X. *Reform in China's agriculture: Trade implications.* EAAU briefing paper series Department of Foreign Affairs and Trade, Australia 9. Barton, ACT, 1997:11–24.
349 Longworth, "China's Pastoral Region", 46.
350 Ibid.:46
351 Ibid.:46
352 Zhizhong, Wu, and Du Wen. "Pastoral Nomad Rights in Inner Mongolia." *Nomadic Peoples* 12, no. 2 (2008): 17. doi:10.3167/np.2008.120202.
353 Ibid.:17

and animal stocking rates gave rise to an ecological crisis demonstrated by the reduction of grassland-plants diversity and a decline in groundwater resources. To stop the ongoing destruction of the grassland ecology, the government introduced a new system described as "Two Rights and One System" (*grassland ownership, grassland use rights, and the household contract responsibility system*) "demarcating pastures, forbidding free nomadism, adopting settled residences, and controlling livestock stocking rates."[354] All of these regulations were implemented under the assumption that the policies would prevent overstocking and the progressive deterioration of Inner Mongolia's ecology. Still the effect of this policy strategy has not caused a stop in deterioration, on the contrary, as Wu Zhizhong and Du Wen (2008) have illustrated.[355] They refer to their questionnaire from Inner Mongolia and point out that the ecology of the grassland has suffered more and the life of herders has become more complicated.[356]

---

354   Ibid.:18
355   Zhizhong and Wen, "Pastoral Nomad Rights in Inner Mongolia", 13–33.
356   Ibid.:19

5. Economy versus Ecology: China's Balancing Act – A Challenge for Inner Mongolia

> *"We won't have a society if we destroy the environment"*
>
> Margaret Mead (1901–1978), Cultural Anthropologist

The following chapter will clarify the issue of national policies in China and their impact on Inner Mongolia specifically. It is not possible to refer to the Inner Mongolian agricultural reforms without considering that they were implemented as a national policy by the Communist Party. Therefore it should be clarified, when describing changes in the national reform policies that they are obligatory for Inner Mongolia. Furthermore, when examining the impacts of rural reforms in China, it is important to know that farmers and herders in Inner Mongolia are both effected by political reforms and both economies are embedded in the domestic economy as major contributors to the GDP of China's primary agricultural sector.

China's dilemma lies in its population density, which at the end of the twentieth century has reached one billion people. With the rising population, agricultural production intensified as well, in an attempt to keep the balance intact. But the total area of the country remained almost the same, which caused and imbalance of food demand and supply. The era of industrialization and technological development under Mao has further increased the pressure on China's ecology. Environmental degradation in China did not only occur because of the industrialization beginning in the early 1950s, but also due to the immense population pressure from the eighteenth century.[357] The intense deforestation along the Yellow River in Ordos and Hetao caused erosion, flooding, and desertification and desolation along whole tracts of land downstream. In summary, China lost more available land than it gained in the previous two centuries. The economy-ecology balance was destroyed long before the industrialization process of 1953 began.[358] In the following decades, as the previous chapters have illustrated, under the government of Mao the country had a tumultuous relationship with the environment. This is marked by the destruction of whole ecosystems. The already

---

357 Glaeser, Bernhard. *Umweltpolitik in China: Modernisierung und Umwelt in Industrie, Landwirtschaft und Energieerzeugung.* Sozialwissenschaftliche Studien, Bd. 20. Bochum: Brockmeyer, 1983:50.
358 Ibid. 51.

mentioned tree-cutting in Inner Mongolia, has resulted in a large-scale transformation from forested slope-hills into cultivated field terraces.

In general, Inner Mongolia was much more affected by in-migration policies than other provinces in China, which has led to the expansion of acreage into the grasslands for extensive cultivation and intensive animal-husbandry. Rivers, like the Yellow River, were dammed and artificial water reservoirs constructed. The "Great Leap Forward" (1958–1961) was largely responsible for the deforestation of huge parts of China to support Mao's aim of transforming China's agrarian economy into a socialist society. This specific deforestation was due to the national steel furnaces, which were established nationwide and required huge amounts of wood to power production. The rapid industrialization and collectivization ended with the Great Famine. The Inner Mongolian grasslands have specifically suffered during the Cultural Revolution (1966–1976) when the rising population demanded grain and the grassland had to be converted into fields for crops.[359] Even Mongol residents, normally sensitive to their ecology, were engaged in transforming the grasslands and gained a positive reputation nationwide for their commitment with Mao's ideology.[360]

To understand the history of environmental degradation in Inner Mongolia and how this formed today's Inner Mongolian environment, it is essential to expand on the political, social and economic contexts in which animal husbandry and agriculture were practiced in northern China since 1949.[361]

5.1 Inner Mongolia as Part of Rural Reforms integrated in China's Agricultural Development

> "We abuse land because we regard it as a commodity belonging to us. When we see land as a community to which we belong, we may begin to use it with love and respect."
>
> Aldo Leopold 1887–1948, Ecologist

Following the foundation of the People's Republic in China, the Chinese government began the implementation of the rural land reforms, which have mobilized

---

359 Jiang, Hong. "Grassland management and views of nature in China since 1949: regional policies and local changes in Uxin Ju, inner Mongolia." *Geoforum* 36, no. 5 (2005): 641–53, 642. doi:10.1016/j.geoforum.2004.10.006.
360 Ibid. 642.
361 National Research Council, "Grasslands and Grassland Sciences", 26.

approximately 70 percent of the population to cultivate 46.6 million hectares of farmland taken and returned from feudal landlords to the people. To organize the mass of people and fulfill the plan of agricultural development in China's rural areas, the communist system transformed the old imperial society into a cooperative and commune-led society, which according to Ellis was based on "the socialist principles of collective ownership [...] and production, collective productive work, and unified (egalitarian) distribution of benefits [and goods]."[362]

Before the land reform in Mongolia began in 1947, the area had been controlled by animal landlords, officers, and monasteries. Their aim was to limit the numbers of livestock from landless herdsmen, but to increase their own herds.[363] Between October 1947 and May 1948, land in the Inner Mongolian region was confiscated by the communist party and redistributed among herders and farmers. The land reform campaign in China was responsible for the death of approximately 10 percent of the Chinese peasants or 1 to 4.5 million people.[364] It is still unclear and debated exactly how many people died during the campaign. From 1947 forward, Chinese agriculture has gone through various phases of collectivization and decollectivization. During the periods of collectivization China was able to keep pace with the growing population and increase its grain production, but economic growth remained constant. In response the government established communes and the former distributed land of herders and farmers was handed over to the village communes by 1958. These communes, to which all community members belonged, were, according to Brown, "the basic production, accounting and administrative unit in rural China."[365] The whole system was aligned to organize labor and distribute the collective income between the people. The problem of the commune system was that the single production teams had very little opportunity to influence their production method and the absence

---

362 Yan, Rui-zhen. "Changes in the system of ownership in rural China." In *China's rural development miracle: With international comparisons papers presented at an international symposium held at Beijing, China, 25–29 October 1987*. Edited by John W. Longworth. St. Lucia, Qld, Portland, Or: University of Queensland Press; Distributed by International Specialized Bk. Services, 1989:11.
363 Rummel, R. J. *China's bloody century: Genocide and mass murder since 1900*. New Brunswick, N.J: Transaction Publishers, 2007:223.
364 Ibid.:221
365 Colin Brown and Chan Kai. "Land reform, household specialization and rural development in China." Agricultural and Natural Resource Economics Discussion Paper 8 Unpublished manuscript, last modified October 16, 2014:2 http://ageconsearch.umn.edu/bitstream/123790/2/BrownChen.pdf.

of economic incentives increased work inefficiency, as the Chinese phrase: *ta hu long* (Everybody's job is nobody's job) explains.[366] Until 1965 there was the period of adjusting the commune system in order to eliminate the emerging problems of limited working capacity, labor quality and work-initiative.[367] The production teams allowed the households to retain some income and eased the restrictions of property rights, which also enabled the cultivation of private plots.[368]

Other factors playing a key role in the difficulties of implementing the collective commune system are described by Yan Rui-zhen in "Changes in the system of ownership in rural China" in 1989[369]:

1. The rural population did not demand a collective organization, therefore there was little work ambition and efficiency, which made the production decrease. The working capacity of peasants never met with the demand of the large-scale agricultural production enforced by the Communist Party. Technology and working tools available in rural China were far from modernization and based on household farming by animal power. Therefore intensive, large-scale crop cultivation was nearly impossible.
2. Before the collectivization the property of the communes consisted only of the means of production possessed by the peasants. The peasants' farmland held 85.7 percent of the total value of property owned by the communes and there was only little accumulation of property after the implementation of the commune system. It took several years to increase well-being and living standards and earn surplus for the peasants working in the collective production.
3. Due to the enormous population involved in the commune system it was impossible to estimate the quality and quantity of labor of any single member. The Communist ideology followed the rule that the distribution of goods and funds was on egalitarian basis. As a consequence, the ambition to work sank because it was commonly believed that it made no difference whether people worked hard or not.[370]

During the era of industrialization the government also traced the agricultural development in rural China. Consistent development of industrialization was difficult because this ambitious plan could not be financed with the low agricultural

---

366 Yan, "Changes in the system of ownership in rural China", 12.
367 Ibid.:12
368 Brown and Kai, "Land Reform, household specialization and rural development", 2.
369 For further information see: 1989 *"China's rural development miracle"*, edited by John W. Longworth.
370 Yan, "Changes in the system of ownership in rural China", 12.

productivity. Therefore the government set the agricultural procurement prices below the market price in an attempt to transfer the resulting surplus in cultivation to industry.[371] What followed were decades of collectivization of agriculture. The collectivization forced farmers maintain production even at low prices, and a major problem of the system was its inefficiency. Because egalitarianism was the underlying ideology of communistic, commune policy, individual work commitment was not rewarded; many people made no serious effort. Therefore when productivity levels of varied systems are compared, it becomes obvious that the productivity of non-commune, i.e. private, plots being cultivated by small units or households are more efficient with five to seven times higher productivity rates than those of the communes.[372] Because personal or private ownership of capital, labor, and land was strictly prohibited during this planning system resource investment was also inefficient. The government thus pushed self-sufficiency in grain production in order to support industrialization. The focus on grain, apart from crop variety and quality, led to a severe food shortage in the 1950s and 1960s. Although the government concentrated on grain production, the per capita output declined and was never able to reach the same peak levels before the takeover of the Communist Party.[373] With the following years of famine during the *Great Leap Forward*, China became a net grain importer. However, to pay for the imports the country was forced to export other foods (mainly grain), which were also short in supply, in exchange.

Despite problems beginning shortly after the program's implementation, the government kept the system in place until 1978. Today it is well known that the commune system could cope with neither the structure of China's rural areas and the economic strength of the people, nor the peasants' ideological consciousness.[374] The farmer did not gain confidence in the commune system because the available working tools were only suitable for subsistence farming by animal power and not for large-scale production. The problem of implementing a social organization in rural areas sparked the people to consider alternatives, which encouraged hard work and eliminated the restrictions enforced by household management systems. This led to the following disastrous years of economic development until 1978.[375]

---

371 Wu, "Reform in China's agriculture: Trade implications", 6.
372 Ibid.:8
373 Ibid.:9
374 Yan, "Changes in the system of ownership in rural China", 11.
375 Ibid.:12

After decades of changing political campaigns exploiting the natural environment in the name of rural development, the reform era under Deng Xiaoping started a program in 1978 to rehabilitate the rural production.[376] What the government did was raise the contract price for grain and decrease the price for pesticides and fertilizers, as well as technical machines, so that the peasants could compete with the market situation.[377] But with these changes the households were able to organize their production with their own labor. The people's commune distributed land to each peasant's household proportionally according to the number of household members or the labor force of the household.[378] Still major problems occurring from 1978 to 1983 included a stagnation of agriculture and productivity, an increase of grain imports in order to secure the food demand of the urban population, a reduction of the living standard in rural areas and a decline in the access to natural resources.[379] Furthermore, the contract prices of grain fell but the cost to maintain large-scale cultivation through the use of fertilizers, pesticides, and technical equipment such as threshing and sowing machines rose.[380] The net-income for farmers stagnated largely because the living costs, including goods and food, increased nationwide due to the immense population growth and therefore the increased demand for products.

The agricultural reform as mentioned before, was later named the *Household Responsibility System* which was widespread over China and affected all economic sectors, namely agriculture, animal husbandry and industry. The HRS led to a rapid increase in the living conditions of the rural population and the gradual development of self-responsibility for farmers. They now were able to decide which crops to plant and how they should be produced. After the payment of agricultural taxes and certain amounts of dues to the cooperatives, the remaining profit from their labors belonged to the household, which increased the farmers' enthusiasm for agricultural production.[381]

Underlying the socialist ideology, the HRS supported the rural development with two mechanisms. First, herders and peasants took part in the rural economy, and were responsible for their production, but also in order to earn a living

---

376 Ibid.:29
377 Grassi, Sergio. "Chinas Agrarreform – in Zeiten der globalen Finanzkrise." http://library.fes.de/pdf-files/bueros/china/05996.pdf.
378 Ibid.:12
379 Grassi "Chinas Agrarreform", 1.
380 Yan, "Changes in the system of ownership in rural China", 19.
381 Yan., "Changes in the system of ownership in rural China", 13.

for themselves.³⁸² Second, the village cooperatives guaranteed a functioning micro-economy for the small-sized households, which were unable to work independently, by supporting water conservation projects, pests and crop disease management or in the distribution of agricultural machines. Although most of the peasants' households disposed of privately-owned agricultural equipment, the contracted land still belonged to the commune, showing the character of the socialist cooperative organization as it was a hybrid between the new system and the old commune system.³⁸³ The economic turn began with this liberalization of agriculture when agricultural reforms transformed the cooperative farming into privately owned household cultivation.³⁸⁴ These changes in policies made agricultural surplus possible, which helped to cope with the Chinese food shortage of past years. In the 1990s, China became a food exporter after decades of chronic shortage in food production. Still, today, China's goal is "grain self-sufficiency" but this can only be accomplished under high costs for the economy, as China lacks arable land and an abundant labor force.³⁸⁵

Under the Reform Era of Deng Xiaoping, China opened for foreign investment and focused on the improvement of the economy as an attempt to leave the stigma of an "undeveloped" country behind. The commune system was abandoned and China slowly transformed from a centrally planned economy towards a socialist market economy.³⁸⁶ This change brought with it better living conditions for most of the people and an increase and improvement in food production. Since the implementation of the rural reforms in 1978, China's agricultural production rose significantly, poverty fell drastically, and China transformed from a country depending on imports to a world leading export country.³⁸⁷ In average China's economic growth recorded 9.5 percent from 1985 to 2005.³⁸⁸

Today the HRS is still in use and farmers are able to lease land-use rights for thirty years. Individual property rights still do not exist. To establish a productive and profit-oriented agriculture under these circumstances is a challenge for the government, farmers, and industries. The most recently implemented reform

---

382 Ibid.:13
383 Ibid.:13
384 Wu, "Reform in China's agriculture: Trade implications", 4.
385 Ibid.:4
386 OECD Review of Agricultural Policies – China. http://dx.doi.org/10.1787/9264012613.
387 OECD Review of Agricultural Policies – China.
388 OECD 2005: *Economic Survey of China*. Foy, Colm and Maddison, Angus, "*China: a world economic leader?*" in *OECD Observer* No. 215 January 1999. www1.oecd.org/publications/observer.

affecting agriculture, animal husbandry, and industry was initiated in 1999, with the *Grain for Green* program. Its major goal is, according to Can Liu and Bin Wu (2010), "to combat deforestation, ecological degradation, over-cultivation on sloping land and soil erosion."[389] Ten years after the implementation the program achieved improvements in environmental ecology, China's forest grass coverage has increased up to 62.9 percent in 2006.[390] The program also led to a reduction in rural poverty and increased the households' incomes, which showed remarkable signs in Inner Mongolia, where the per capita GDP exceeded the national standard by 2004.[391] The political achievements since the reform era are remarkable. Improving rural markets through reforms by giving the small-farm households the chance to rent land and enter the local market proved essential to stimulating the national economy. But China's insecure property and land-use rights as well as legal restrictions of land-use policies have also negatively affected the smallholders' access to development. Therefore promoting irrigation and improving water-access for the rural poor was essential. Development aid in terms of education and technical support were also crucial for rural development and economic growth. Additionally, the guarantee that the rural population gained access to health care was of great importance. But the fast economic development increased China's environmental problems, as the historical review shows.

5.2 Economic and Environmental Background of China

From 1952–1978, China's per capita GDP raised roughly 3 percent per year. This growth rate was predominantly due to increases in government investments as well as a surge in the people's education level.[392] But the enormous disruptions of the Great Famine and the Cultural Revolution led to a regression in China's productivity. However, the country's economic growth again showed

---

389 Liu, Can, and Bin Wu. "Grain for Green Programm in China: Policy Making and Implementation?" Briefing Series, China Policy Institute, University of Nottingham, April 2010. https://www.nottingham.ac.uk/cpi/documents/briefings/briefing-60-reforestation.pdf.
390 Wang, Ji-Jun; Jiang, Zhi-De; Xia, Zi-Lan 2014:"*Grain-for-Green Policy and Its Achievements*". In: Atsushi Tsunekawa, Guobin Liu, Norikazu Yamanaka und Sheng Du (Hg.): Restoration and Development of the Degraded Loess Plateau, China: Springer Japan (Ecological Research Monographs), p. 141.
391 Ibid.:256
392 Zhu Xiaodong 2012: "Understanding China's Growth: Past, Present and Future. Past, Present, and Future". In: The Journal of Economic Perspective 26 (4), p. 103–124, 103.

impressive development since the reform era of 1978. The former political concept of economic separation under Mao Zedong's (毛泽东 1893–1976) leadership has transformed into integration into the world economy with the "Reform and Opening Up" led by Deng Xiaoping (邓小平 1904–1997). China's economic development has been marked by steady growth, which was on average, has been more than 10 percent per year. Since 1978, the reforms have boosted China's annual industrial output rate by more than 11.4 percent.[393] 1997 marked a milestone in China's further economic development. The Fifteenth Congress of the Communist party declared the new policy of trade liberalization and privatization.[394] Together with the legalization of private enterprises, the reduction of legal barriers, the extension of trade rights and the opening for foreign direct investment, China also joined the World Trade Organization in 2001.[395] Between 1998 and 2007, the annual total factor productivity growth rate of state and non-state sector was 3.61 percent. In 2007 China's GDP has already passed Germany's GDP with an annual growth rate of 13 percent, made it at the time the world's third largest economy. Since 2009 China is ranked second worldwide.[396]

---

393 Council on Foreign Relations *"China's Environmental Crisis"*: http://www.cfr.org/china/chinas-environmental-crisis/p12608.
394 Zhu, "Understanding China's Growth", 117.
395 Ibid.:117
396 Peter E. Robertson. "The Global Impact of China's Growth: Economics." In *The Oxford companion to the economics of China*. Edited by Shenggen Fan. 1. ed. Oxford: Oxford University Press, 2014:98. Accessed February 8, 2016. http://ndc.gov.bd/lib_mgmt/webroot/earticle/611/The-Global-Impact-of-Chinas-Growth.pdf.

Fig. 4: Development of China's Economy

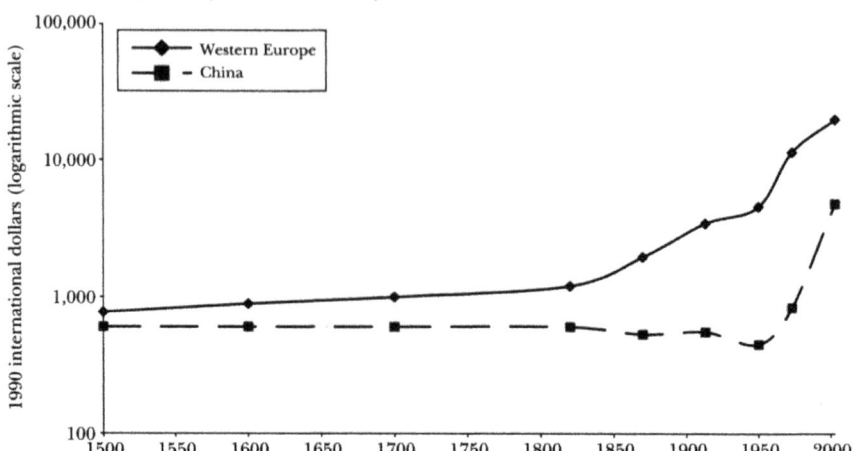

Source: Madison, Angus. 2007. "*Contours of the World Economy, 1–2030 AD*". Oxford and New York: Oxford University Press.

Despite all its economic successes, China still demonstrates elements of a developing country. Although absolute poverty has declined from 835 million people in 1981 to 207 million in 2005,[397] 470 million Chinese still have an income below two dollars per day,[398] thus ranking China second after India in terms of people living in poverty. At the end of 2012, 98.99 million people still lived below the national poverty line of 2,300 RMB (*Yuán* 元) per year.[399] China's economic status in the remote and rural areas, the low living-standard for the majority of people as well as the large development gap within different regions and provinces show that China is fully in the "developing process".[400] But in relation to economic growth, it is incorrect to label China with the status of a "developing country" due to the fact that since 2014 China has surpassed the US in terms of economic

---

397 China Daily "*China set to raise poverty line*": http://www.chinadaily.com.cn/china/2008-09/03/content_6992004.htm.
398 Rudolph, Jörg, and Thomas Heberer. *China – Politik, Wirtschaft und Gesellschaft: Zwei alternative Sichten*. Sonderausg. für die Zentralen für politische Bildung in Deutschland. Forum hlz. Wiesbaden: Hessische Landeszentrale für Politische Bildung, 2010:122.
399 http://www.worldbank.org/en/country/china/overview.
400 http://www.worldbank.org/en/country/china/overview.

power. Nevertheless, with regard to environmental problems, the fast economic growth is responsible for worsening the environmental conditions.

Increases in economic growth often involve rapid development at the cost of the environment. Facing this huge challenge of how to grow economically without exploiting natural resources, starting in 2011, China implemented its latest *Five-Year Plan* (2011–2015),[401] emphasizing the rebalancing of its economy and environment. Main targets of the Twelfth Five-Year Plan for resource conservation and environmental protection are:

> [China] will maintain farmland reserves at 1.818 billion mu (approximately 121,260,600 hectares). [China] will cut water consumption per unit of value-added industrial output by 30%, and increase the water efficiency coefficient in agricultural irrigation to 0.53. Non-fossil fuel resources will rise to 11.4% of primary energy consumption. Energy consumption per unit of GDP will decrease 16% and CO_2 emissions per unit of GDP will decrease 17%. [China] will make significant reductions in the total emissions of major pollutants: chemical oxygen demand (COD) and SO_2 by 8%, ammonia nitrogen and nitrogen oxide by 10%. Forest coverage rate will increase to 21.66% and national forest stocks will increase by 600 million cubic meters.[402]

The broader goal of this new plan is to reach sustainable growth, stimulate domestic consumption and improve the social infrastructure. China will transform from a producer culture to a consumer society, which means that economic growth is guaranteed through stimulated domestic consumption, boosting employment, wage increases, and shifting the disparity of income away from saving towards spending. It is important to transfer the development of former coastal and urban areas to the remote rural regions in order to eliminate economic and social disparity. Therefore the inland areas must become more hospitable to growth and development, thus it will be necessary to strengthen the domestic economy, leave the environment intact and improve social infrastructures.[403]

The government's first priority is to alleviate poverty and social disparity as well as environmental degradation to create a homogenous basis for economic development.[404] However, overall, China continues to focus on productivity

---

401 http://www.britishchamber.cn/content/chinas-twelfth-five-year-plan-2011-2015-full-english-version. For details, see full document in Chinese at: http://www.moa.gov.cn/ztzl/shierwu/hyfz/201112/t20111227_2444181.htm.
402 Ibid.:3,4.
403 http://www.britishchamber.cn/content/chinas-twelfth-five-year-plan-2011-2015-full-english-version.
404 "China's Twelfth Five-Year Plan (2011–2015) forcefully addresses these issues. It highlights the development of services and measures to address environmental and

increases and yield rise for farmers in order to be able to both compete economically with other nations and meet its own needs. Therefore governmental policies place emphasis on the rural areas, where China's structural heterogeneity has a major impact on the development dynamics. Whereas urban regions are highly developed, the majority of rural China is still lagging behind in terms of infrastructure, access to medical support, education and average income.[405] While big cities attract rural labor, large tracts of agricultural land are left unused as a result of urbanization. Agricultural areas are farmed inefficiently, due to the lack of technology and the inability to adopt more current technology.[406] The interregional inequality in economy, geography and demography exacerbates the difficulties associated with maintaining a consistent economic policy and a uniform accrual rate. The Twelfth Five-Year Plan highlights the formidable challenges of China's further development:

> "However, it is important to have a clear sight of the imbalanced, incompatible and non-sustainable elements within China's development, which mainly turn out to be a tightened constraint between economic growth on one hand and resources and environment on the other, an imbalance between investment and consumption, a relatively large income disparity, uncompetitive technological innovation ability, unreasonable industrial structure, vulnerable agricultural basis, a gap between rural and urban development, a coexistence of total employment pressure and structural contradiction, a significant increase in social conflicts and a still considerable number of institutional obstacles that restrain scientific development."

Seventy percent of China's rural population has gained only a primary or middle school education. This generation's children usually leave the countryside and move to cities, for better job opportunities or attain higher levels of education. Rural China is thus confronted with emigration to the cities and affected by an increase in "brain drain".[407] Additionally, every province is marked by different factors which influence the economic, social, and environmental development.

---

      social imbalances, setting targets to reduce pollution, to increase energy efficiency, to improve access to education and healthcare, and to expand social protection. Its annual growth target of 7 percent signals the intention to focus on quality of life, rather than pace of growth", available at: http://www.worldbank.org/en/country/china/overview.
405 http://www.britishchamber.cn/content/chinas-twelfth-five-year-plan-2011-2015-full-english-version, 4.
406 Gang Chen. "The seeds of China's future." *The World Today* February & March 2015 (2015): 36–39:36. Accessed July 21, 2015.
407 Ibid.:36

Population density in the south-eastern provinces is much higher than in the north-west.[408] China has a dual-structured urban and rural economy: A rural resident in 2013 earned approximately 8,895 RMB per year, whereas an urban resident earned 26,995 RMB per year.[409] Considering the environmental factors, the geographic position of each area may influence the hard and soft location factors of industry. The northern regions, especially Inner Mongolia with its natural deposits of coal and oil, are as a matter of course interesting for the mining industries. In contrast, the fertile and humid regions of South China, namely the Middle Lower Yangtze Plain and the Pearl River Delta are perfectly suited for paddy rice field cultivation.[410]

Chinas demographic structures also have a strong bearing on its economic development. As such, the composition of the population is directly affecting the labor force, especially as it relates to population aging, gender imbalance, and life expectancy.[411] Here Chinas ethnic potpourri has influenced the socio-economic development throughout history. Beyond the obvious distinctions between male and female, adults and children, China's provinces are typically characterized by a different ethnic composition of their population. Fifty-five ethnic minorities live in China, each with different attitudes towards politics, the economy and culture, as well as unique understandings of the environment, which are deeply rooted in each culture. The North and the South are distinguishable from each other in their varied economies, eating habits and religious or spiritual beliefs as well as the impact of differing economic development under the Chinese Communist Party since 1949. Historically, the South was inhabited by traditional tillers cultivating rice using Asiatic water buffalos and bovine, whereas the North was populated by nomads, raising livestock like horses, sheep, goats and pigs. The dietary habits were connected to the living conditions as the South

---

408 Ibid.:39
409 Ibid.:36
410 Xiao, Xiangming, Stephen Boles, Jiyuan Liu, Dafang Zhuang, Steve Frolking, Changsheng Li, William Salas, and Berrien Moore. "Mapping paddy rice agriculture in southern China using multi-temporal MODIS images." *Remote Sensing of Environment* 95, no. 4 (2005): 480–92. Accessed July 21, 2015. doi:10.1016/j.rse.2004.12.009.
411 Eggleston, Karen. "China's demographic change in comparative perspective: Implications for labor markets and sustainable development." *Re-evaluating labor market dynamics a symposium sponsored by the Federal Reserve Bank of Kansas City, Jackson Hole, Wyo., Aug. 21–23, 2014*, 2015, 203–31.

consumed mainly rice, and the Northern people instead ate millet and wheat.[412] Today's eating habits are still based on this traditional division, but most of the cultural habits and traditions have since been Sinicized and influenced by the Han-Chinese society. In terms of religion or spirituality, the Southern Han population was characterized by Confucianism and the associated ethics and moral concepts of a holistic harmony that was banned under Mao and replaced by the belief of man's dominion over nature.[413] In contrast, the Northern nomadic tribes practiced Animism and Shamanism, which are both characterized by the belief that every being possesses a sort of spirit or soul. This deeply rooted tradition could however be maintained even after the introduction of Buddhism under Althan Khan as well as under Communist governance because the idea of the spirit of nature could not be easily suppressed. Since the founding of the Peoples Republic in 1949 China itself claims to be a laicistic state and the people officially benefit of religious liberty. The government's attitude towards religion is still ambivalent, but the mixture of Confucianism-Daoism-Lamaism is accepted and commonly practiced by the people.

Besides demography, the historical in-migration of ethnic Han has changed many aspects of the former Mongolian society as Rong Ma, author of "Changes in the nomadic pattern and its impact on the Inner Mongolian steppe grasslands ecosystem", has pointed out.[414] Population density, ethnic structure, economic structure, social organization, language use, life customs, and the ecological environment have all been affected by this migration.[415] Although the majority of the Mongol population seems to have assimilated into the Chinese society, life in Inner Mongolia's public sphere still shows all signs of a multi-ethnic society.

---

412 张宏洲 2015:土豆：缓解粮食、污染、干旱危机的灵丹妙药？
 Zhang Hongzhou 2015: "Can the potato help feed China, cut pollution and alleviate drought?" https://www.chinadialogue.net/article/show/single/ch/7657-Can-the-potato-help-feed-China-cut-pollution-and-alleviate-drought-
 土豆的主粮化也会有助于校正农业生产中日益严重的南北失衡.几百年来气候宜人、水量充足的南方一直是中国粮食生产的重心所在，但如今这个重心正不断向干热的北方转移，所以必须耗资800亿美元修建南水北调工程.
413 Neumann, Cilia 2011: "Aschenputtels Erbe: Lilienfüße in China" (Magisterarbeit): In der antiken chinesischen Gesellschaft diente der Konfuzianismus dazu, soziale Hierarchien und das patriarchale Clan-System zu stärken, Regeln für Gesetze und Riten zu konsolidieren und moralische und ethische Standards festzusetzen […], 39.
414 Ou Li; Rong Ma; James R. Simpson "Changes in the nomadic pattern", 63–72.
415 Rong, Ma. "Migrant and Ethnic Integration in the Process of Socio-economic Change in Inner Mongolia: a Village Study." *Nomadic Peoples*, no. 33 (1993): 173–91. Accessed January 9, 2015.

The Mongolian language is omnipresent. Bilingual street signs, commercial outlets, and government documents are required to be provided in Chinese and Mongolian. In Hohhot, Mongolian schools and kindergartens for the Mongolian-speaking population are common, and both television and radio programs are also available in both languages. Indeed, Rong Ma has already confirmed during his research in 1993 that despite the different policies of cooperative transformation, the commune system and the HRS, even during the Cultural Revolution, there was little ethnic conflict or conflict of interests between Han and Mongols and they have lived harmoniously since the 1930s.[416]

In terms of economic development, South China continues to have an environmental and ecological advantage when compared to North China. This has been further complemented and strengthened by the development of a commercialized rural agriculture and the establishment of an urban consumer-goods industry.[417] With economic improvements under the Chinese Communist Party and Deng Xiaoping's new market-oriented reform, China's coastal areas developed at a rapid pace. Today the former nationalistic South is much more commercially oriented and still utilizes of the special economic zones[418] (SEZ). The North, where Mao's triumph began, still carries the burden of the early years of the Communist rise and continues to focus on agriculture, livestock, and heavy industries, like coal and rare earth.[419] Although the southern population also lives under rural conditions, their knowledge of farming and the ability to share information appears to have been passed on over generations.[420] A sense of responsibility to the environment and a strong connection with the idea of homeland may be a reason for a much more pronounced investment and interest in higher yields and better profitability.

China's economy has demonstrated fast growth as have environmental problems in the country. As the world's largest electricity supplier and producer of

---

416 Ibid.:189
417 Lin, G.C.S. *Red Capitalism in South China: Growth and Development of the Pearl River Delta*. Vancouver: UBC Press, 2011:63.
418 The special economic zones are characterized by specific economic policies and flexible governmental guidelines, like tax incentives for foreign investment and greater independence on international trade activities.
    Leong, CheeKian. "Special economic zones and growth in China and India: an empirical investigation: International Economics and Economic Policy." *Int Econ Econ Policy* 10, no. 4 (2013): 549–67. doi:10.1007/s10368-012-0223-6.
419 Ibid.:65
420 Ibid.:64

steel, cement, aqua cultured seafood, and chemical textiles, China's demand on resources has been non-stop.[421] In particular the production of pesticides for agriculture has continuously grown over recent decades. The use of fertilizers and pesticides in Chinese history can be dated back to the Ming Dynasty as they have played an important role for the successful development of agriculture.[422] Indeed, today it is difficult to imagine a productive large-scale agricultural system without the use of pesticides. China is no exception in this as Zhang Wenjun notes in "Global Pesticide Consumption and Pollution: With China as a Focus," stating that "China has become the largest pesticide producer and exporter in the world."[423]

As this thesis is dealing with environmental problems related to dryland agriculture and crop production in fragile ecosystems, a short discourse on pesticides, their effect and impact will be added here. Pesticides are any substances utilized for preventing, destroying, repelling or mitigating insects, mites, nematodes, weeds and rodents. This includes insecticides, herbicides, fungicides among diverse other substances to control pests.[424] Without the use of pesticides the estimated crop loss for fruits, vegetables and cereals would be 78 percent, 54 percent and 32 percent worldwide, respectively.[425] Comparable projections can be found in the *Chinese Yearbook on Agriculture 2007*, where it was estimated that a loss of approximately 90 million tons of cereals, 1.7 million tons of cotton, 2.6 million tons of oilseed and 80 million tons of vegetables would occur without the application of pesticides.[426] Undoubtedly, most pesticides are potentially harmful, as they are specifically produced to kill animals, pests or other living organisms. However, chemical crop protection remains essential to control insects,

---

421   Liu, Jianguo, and Jared Diamond. "China's environment in a globalizing world." *Nature* 435, no. 7046 (2005): 1179–86. doi:10.1038/4351179a.
422   "China is one of the earliest countries to use pesticides. As early as in Ming Dynasty, the monograph Ben Cao Gang Mu, edited by Li Shizhen, has recorded a number of plants and minerals that used as pesticides such as veratridine, flavescens, arsenolite, realgar, orpiment and lime, etc, (Chen, 2007)" Zhang 2011:130.
423   Wenjun Zhang, Fubin Jiang, Jianfeng Ou. "Global pesticide consumption and pollution: with China as a focus." *International Academy of Ecology and Environmental Sciences* 1, no. 2 (2011): 125–44. Accessed January 11, 2015.
424   EPA 2015: http://www.epa.gov/pesticides/about/index.htm#what_pesticide.
425   Cai D.W. "Understand the role of chemical pesticides and prevent misuses of pesticides." *Bulletin of Agricultural Science and Technology*, 36–38 (2008): 36–38.
426   Editorial Board of Chinese Yearbook on Agriculture, 2007.

diseases, weeds, fungi and other undesirable pests.[427] As *CropLife*, an international non-profit trade association of agribusiness, explained:

> Food crops must compete with 30,000 species of weeds, 3,000 species of nematodes and 10,000 species of plant-eating insects. We know that despite the use of modern crop protection products 20–40% of potential food production is still lost every year to pests. These losses can occur while the crop is growing in the field, when it is in storage and in the home. In short, an adequate, reliable food supply cannot be guaranteed without the use of crop protection products.[428]

Against the backdrop of an ever-increasing world population and the growing need for a guaranteed food supply the application of pesticides should be continued. This is especially true for China, which is confronted with the particular problem of land loss, thus the application of pesticides is absolutely essential to guarantee the best possible yields. Still the transition towards organically-based and bio-degradable pesticides, such as pheromones and microbial pesticides should be considered, as they are often safer than more standard products. Reflecting the impact of environmental action groups who have been pushing for safer and more sustainable pesticide use, the Environmental Protection Agency is already registering "reduced-risk conventional pesticides in increasing numbers."[429]

With increasing wealth, other agriculture sectors in China have also grown. For example, both the meat production and dairy industries have grown fourfold between 1978 and 2002.[430] The problem of this rapid growth in China's production means that greater amounts of agricultural waste is being left behind as a by-product and this is impacting the environment and ecology. When looking at the industrial sector such as coal-mining or chemical production it becomes clear that the technology is often outdated. These outdated processes and methods further increases environmental pollution and therefore stress on ecosystems. The increased production and consumption of both industrial and agricultural products have led to a surfeit of both organic and non-biodegradable waste, which has added to China's current environmental crisis connected to an ever-increasing amount of air, water and land pollution.

---

427 CropLife 2015. Available at: https://croplife.org/crop-protection/benefits/ Accessed January 2016.
428 Ibid.
429 EPA 2015. Available at: http://www.epa.gov/pesticides/about/index.htm#balance. Accessed January 2016.
430 Liu Jianguo, Diamond Jared "China's environment in a globalizing world", 1180.

5.3 Socio-Economic Situation of Inner Mongolia

Marked by the geographical transition line of fertile pasturelands in the North and the large agricultural areas alongside the Yellow River, it was already mentioned, that Inner Mongolia was predestined for in-migration. But today Inner Mongolia's economic development as well as social welfare is threatened by environmental destruction, the begin of which can be dated back to the sudden population increase in the mid-twentieth century. Inner Mongolia's ecological issues, which are mainly overgrazed grasslands, desertification and soil erosion are related to human-induced processes due to poor farming practices. The Han population of Inner Mongolia increased from 1.2 million in 1912 to 17.3 million in 1990.[431] By 1997, population pressure and struggles for political reform since the 1960s had pushed Inner Mongolia into the corner as the poorest province of China as measured by disposable per capita income for both urban and rural residents. But by 2012, the GDP per capita in Inner Mongolia passed the benchmark of $10,000 (US) for the first time.[432] Inner Mongolia demonstrated a GDP growth rate of the nearly 12 percent.[433] Thus it is clear that Inner Mongolia has also benefitted from the economic boom that China is currently experiencing.

Although Inner Mongolia's primary economic sector (agriculture) is being diminished for the benefit of secondary (production of goods) and tertiary sectors (production of services), agriculture and livestock production, including cashmere, dairy products and meat are still strong, accounting for 9.1 percent of Inner Mongolia's GDP in 2012[434]. In the more arid grasslands the herding of goats and sheep is advantageous and still remains a traditional method for sustaining a subsistence economy. The large number of sheep and goats further indicate Inner Mongolia as the leading production base for sheep's wool and cashmere, accounting for 26 percent and 42 percent of the agricultural sector 2012, respectively. In terms of dietary specialties, Inner Mongolia has the largest

---

431 Rong, Ma, "Migrant and Ethnic Integration in the Process of Socio-economic Change in Inner Mongolia", 173.
432 China's Provinicial GDP figures. Available at: http://www.china-briefing.com/news/2013/05/16/chinas-provincial-gdp-figures-in-2012.html. Accessed January 2016.
433 Top 10 regions in China. Available at: http://www.china.org.cn/top10/2013-11/13/content_30587616.htm. Accessed January 2016.
434 Wang Hao, Chen Jia and Yang Fang 2013: Coal-rich region looks to 'clouds' for growth. In: China Daily. Available at: http://usa.chinadaily.com.cn/epaper/2013-02/07/content_16213044.htm.

output of mutton, which accounted for more than 22 percent of the national total in 2012.[435]

Tab. 8: Composition of Inner Mongolia's GDP

GDP	2000	2012
Primary	22.8	9.1
Secondary	37.9	55.4
Industry	31.5	48.7
Tertiary	39.3	35.5

Source: Inner Mongolia Statistical Yearbook, 2013[436]

In agricultural cultivation corn remains the leading grain crop, and the farming of grain crops is mainly found along the river valleys and in the fertile areas of Ordos. Other industrial crops include oil-producing crops like sunflower and vegetables but also fruits including tomatoes and melons. Vegetables are predominantly cultivated in the Hetao Irrigation District. In recent years Inner Mongolia has become a leading production base for potatoes in China as it is actively being supported by the national government starting in 2015. Wineries in China are becoming more and more popular. Thus a relatively new economic sector is the cultivation of wines and growing of grapes in Inner Mongolia. Winemaking has become an important economic player in the Wuhai area. This can be connected to changing Chinese dietary habits as drinking wine has become more popular, especially as it relates to the import of foreign goods and lifestyle products as well as cultural transfers. The western influence in China is omnipresent: fashion, food, media and technology from European and American companies are favored by the new generation. Thus because Wuhai's soil structure is suitable for the cultivation of grapes, the region now advertises itself as the "most northern wine-region" of China, fitting the new consumer demands.

Beside the agricultural sector, which includes crop cultivation as well as meat and dairy production, Inner Mongolian economic growth has largely been attributed to natural brown coal resources and huge deposits of rare earth metals.

---

435   Australia China Business Council: http://acbcnetwork.com.au/china-business-tips/know-your-china-region/north-china-including-beijing-hebei-shanxi-tianjin-inner-mongolia/.
436   Inner Mongolia market profile. Available at: http://china-trade-research.hktdc.com/business-news/article/Fast-Facts/Inner-Mongolia-Market-Profile/ff/en/1/1X000000/1X07T7RO.htm. Accessed January 2016.

Approximately one-quarter of the world's total coal reserves come from Inner Mongolia, and in terms of rare-earth metals, China holds more than 97 percent of the worldwide rare-earth resources.[437] But the mining and production processes of rare-earth metals have caused severe environmental damages when not carefully controlled.[438] The production residues of rare-earth metals are extremely dangerous as they are toxic at exposure.[439] The rare-earth mining industry in Baotou, Inner Mongolia is particularly responsible for the uncontrolled dumping of waste, contaminating rivers and farmlands. In 2005, a case of radioactive contamination in the Baotou area and the Yellow River was connected to this uncontrolled waste dumping. Cindy Hurst, analyst for the US Army's Foreign Military Studies Office, reported that, "In the Yellow River, in Baotou, the fish all died. They dump the waste – the chemicals into the river. You cannot eat the fish because they are polluted. Some 150 million people depend on the river as their primary source of water."[440] Thus the ecological destruction associated with mining is a severe problem in Inner Mongolia, especially because of the prominent role the industry plays in the development of Inner Mongolia's economic growth and in the national export economy. The areas around where the mining is done are contaminated as well as the groundwater and the surrounding water environment. Furthermore, the people work in the mines with low safety standards and hence they often suffer from lung diseases and cancer caused by the radioactive elements, as surveys from Baotou have shown.[441] The dilemma is that although Inner Mongolia's economic growth can largely be attributed to its natural deposits of ore, coal, and rare-earth metals, its reliance on energy resources to maintain these industries and the growing population demonstrates well the

---

437 Cindy Hurst. "China's Rare Earth Elements Industry: What can the West Learn.", 3. Accessed June 11, 2013. http://www.iags.org/rareearth0310hurst.pdf.
438 Ibid.:5
439 Cindy Hurst. "The Metal's Edge: The Rare Earth Dilemma: China's Rare Earth Environmental and Safety Nightmare." 2010. Accessed August 10, 2015. http://fmso.leavenworth.army.mil/documents/China's-rare-earth-nightmare.pdf.
"Every ton of rare earth produced, generates approximately 8.5 kilograms (18.7lbs) of fluorine and 13 kilograms (28.7 lbs) of dust; and using concentrated sulfuric acid high temperature calcination techniques to produce approximately one ton of calcined rare earth ore generates 9,600 to 12,000 cubic meters (339,021 to 423,776 cubic feet) of waste gas containing dust concentrate, hydrofluoric acid, sulfur dioxide, and sulfuric acid, approximately 75 cubic meters (2,649 cubic feet) of acidic wastewater, and about one ton of radioactive waste residue (containing water).", p. 16.
440 Cindy Hurst, "China's Rare Earth Elements Industry: What can the West Learn", 17.
441 Cindy Hurst, "The Metal's Edge: The Rare Earth Dilemma", 4.

expected development of Kuznets' Curve[442]. Thus with improved technology and more reliable energy sources the general well-being of the population increases, and these increases in status allow for a new type of reflection on the impact that these industries have on the environment. This newer dialogue then typically results in changes of policy or public thinking about how to move forward with these technologies like rare-earth mining, in a way which that less damaging to the environment and more sustainable. As long as mining is profitable and due to the fact that there are few opportunities to store energy from renewable sources, China has been forced to rely on a coal-based energy supply thus leading to increasing environmental pollution. Therefore the development of a sustainable energy supply is very important for the further economic and environmental development of Inner Mongolia. A future-oriented renewable opportunity lies in the wind power capacity of the high plateaus, where some large wind mills have already been built in 2009.[443] These sectors of renewable energy are interesting to international companies, gaining the attention and participation of these potential and actual investors, and increasing the potential for new support for the economic development of the region.

---

442 See chapter 1.4.2.
443 Kyushu Electric Power Co. Inc.:China Inner Mongolian Wind Farm Project. Available at: http://www.kyuden.co.jp/en_overseas_ipp_ipp07.html. Accessed February, 1, 2016.

6. Empirical Research I: Intensive Cropping in Inner Mongolia: Potato and Poverty Alleviation

This chapter is part of the field research conducted in Inner Mongolia, where the national potato production was chosen as a typical example visualizing the difficulties of Inner Mongolia's agriculture and the contradictory interest of economic growth and sustainable agriculture in a region characterized by environmental distress. First the strength of the potato's contribution to poverty alleviation as a staple crop will be introduced together with a historical outline on the potato's role in the region. Then the national potato policy of China will be addressed. Finally, the particular potato production in Siziwang Qi Banner, Inner Mongolia, which resembles a typical region of potato cultivation dominated by smallholders will be discussed. The evaluated results of the interviews with Inner Mongolian potato producers will provide detailed insight into the business, which has had to and will continue to face the challenges of economic, environmental, and global competitiveness.

The disproportionately population increase (five to six-fold) between the fourteenth century and the early nineteenth century,[444] let Chinese farmers raise their grain output over these six centuries by expanding the acreage and by increasing the yield per acre. But by the nineteenth century China had already run out of cultivable land, and only recorded an increase of 40 percent land area prior to 1957 due to the land acquisition of poor quality in Manchuria, Inner Mongolia and elsewhere in the north-west.[445] The north-western regions continuously expanded their cultivated acreage with the resettlement of the dry land of the north-west.[446] Although the richest lands of China have historically been the rice fields alongside the Yangtze River basin in the mid-south. But the southern areas already began to show a decline between 1400 and the twentieth century during Ming and Qing Dynasty, mainly due to climatic changes, with a shortage of precipitation, which caused a shortage in acreage in the south, thus stimulating the northern Chinese dryland expansion.

---

444 Perkins, Dwight H., "Agricultural Development in China 1368–1968", 13.
445 Ibid.:27
446 Ibid.:18

The increase of Han-Chinese immigration into the northern peripheries and the demand for food rose and the areas for cultivation were also forced to expand, with the exception of the area of HID, where vegetables and other cash crops were being cultivated. The northern agricultural regions of Inner Mongolia were not suitable for vegetable cultivation and irrigation-intensive plants. The raw climate, as described previously, was and remains a challenge for any plant beyond grassland vegetation and bushes.

Today the Chinese government is fully aware of the ecological weakness of Inner Mongolia and the economic strength of the potato. Their solution to solve the gap between economic demands and Inner Mongolia's underperforming agriculture is pushing the domestic potato production. The Chinese Ministry of Agriculture predicts that by 2025, 50 percent of the national potato production will be for the domestic market.[447] Even China's vice Minister, Yu Xinrong, expects a huge advantage from encouraging the potato production in China, commenting that "[the] potato can survive in cold, drought and [a] barren environment. It has great potential to be planted in large vacant fields in the south during winter."[448] Thus, the potato, according to the government, has the potential to serve as a cold-weather crop and fill the gap of fallowness which typically occurs during the winter. It is North of China in particular, however, with its huge plains and parched lands that is the focus of the government's goal to manage the requirements of an increasing food supply. Potatoes are less water-intensive than rice or wheat, and in some areas of the arid north (Inner Mongolia, Gansu and Heilongjiang) the potato is the only crop that grows. Therefore the government focuses on non-endemic crops, which can easily adapt to the natural climatic conditions and are more pest-resistant.

But the cultivation of these new potato crops is accompanied by immense additional effort and the Chinese government relies on technological support and investment of crop specialists. One actor in the field of crop protection is the Swiss company *Syngenta*, which invests in the potato's rising business in China and has developed a project to strengthen the national potato production and yields.[449] The project "Potato Healthy Tubers" (Ma Ling Long 马铃龙)

---

447 *"China to position potato as staple food"* Available on: CCTV.com 01–08–2015 04:13 BJT. Last access February 1, 2016. http://english.cntv.cn/2015/01/08/VIDE1420661517562206.shtml.
448 Ibid.
449 Te-Ping Chen, Chuin-wei Yap, "China pushing potatoes", published online January 12, 2015. Available at: http://www.potatogrower.com/2015/01/china-pushing-potatoes. Accessed Feburary 2, 2016.

was initiated by two former and one current employees of Syngenta and developed in 2011.[450] The project was launched at the World Potato Congress, held in July 2015 in Beijing.[451] This project also became part of *Syngenta's Good Growth Plan*.[452] The plans objectives are six commitments, which are: *"Make crops more efficient, Rescue more farmland, Help biodiversity flourish; Empower smallholders, Help people stay safe and Look after every worker".*[453] For Inner Mongolia specifically Syngenta's action is to help smallholders closing their yield gap and improving land productivity. "Chinese potato farmer don't have sufficient access to healthy tuber seeds", as Hongzhi Zhang, from Syngenta points out. The potatoes, which are kept as seed tubers are "either too big or too small to be sold" in the local markets.[454] The problems are that diseases or pests which are inherent in the plant are passed on to the next generation and the infection can be responsible for a yield drop of 50%.[455] What Syngenta offers the farmers is supply with healthy seed tubers, which were treated with Syngenta products to be more resistant to potato diseases. Syngenta additionally provides post-harvest services for the farmers, like access to machinery and to companies in the potato value chain like McCain.[456] Syngenta further trains thousands of farmers in safe and effective use of pesticides and sustainable farming practices.[457]

The Introduction of the Potato in China

The origin of the potato lies in the Andes of South America. The indigenous Inca population cultivated the *papa* as early as 7000 years ago, establishing a food preservation process of freezing and drying the tubers for longer-term

---

450 Dr. Christoph Neumann, David Liu and Camilo Villalobos.
451 China Potato Expo 2016 in Kunming, China. Available at: http://www.chinapotato expo.com/index.php/en/. Accessed February 1, 2016.
452 Syngenta, Good Growth Plan. Available at: http://www.syngenta.com/global/corpo rate/de/goodgrowthplan/home/Seiten/homepage.aspx. Accessed February 1, 2016.
453 Syngenta, Good Growth Plan. Available at: http://www.syngenta.com/global/corporate/de/goodgrowthplan/commitments/Seiten/commitments.asp. Accessed February 1, 2016.
454 Hongzhi Zhang, "The Chinese Year of the Potato", *Science Matters* Autumn 2013:12. Available at: http://www.nxtbook.com/syngenta/Syngenta_Research_and_Develop ment/Autumn_2013/index.php?startid=13.
455 Ibid.:12
456 Ibid.:13
457 Syngenta, Good Growth Plan. Available at: http://www.syngenta.com/global/corpo rate/de/goodgrowthplan/commitments/Seiten/help-people-stay-safe.aspx. Accessed February 1, 2016.

storage.[458] Spanish conquerors recognized that the potato was of inestimable value and brought samples back to Spain to cultivate in Europe. The potato was introduced to Europe in the late sixteenth century. From Spain, the potato spread to Italy and Germany. Dutch settlers then introduced the potato to Asia via Taiwan (1624–1662). It was around this same time that it also reached China.[459]

Although the cultivation of potatoes in China dates back only 400 years, over the past few decades it has become a crop of enormous importance to China.[460] Production per capita has nearly tripled since the early 1960s, and the cultivation of the "Golden Tuber" is being pushed even further for the future.[461] Due to its short growing period and the ability to adapt well to local, natural conditions like soils and seasonal weather patterns, the potato plays an important role all over the world for food security, energy security, and poverty elimination.[462] An additional aspect is the fact that natural disasters like droughts and floods are often followed by poor harvests for most of the crops in the affected area. Furthermore, it is even more difficult to use the sprouted seeds of the destroyed crops for the new harvest. Because the potato is a tuber plant, it is less affected in such situations and can be transplanted even after a natural disaster, when water covers the fields. In contrast, wheat, corn, and rice are typically rotten after flooding, and are thus less adaptable in such situations when compared to potatoes. During the 1998 China floods, which was a series of floods in the Yangtze River Basin, caused by tremendous rainfalls from July to September, many rice fields in Sichuan were destroyed, but the planting of potatoes and sweet potatoes helped to secure the potential food-shortage crisis in the following season.[463] Of the benefits of increasing potato crops the most significant are its high yield

---

458 Stamp, Hans P. ... *und weiss wie Alabaster: Eine Kulturgeschichte der Kartoffel*. Neumünster: Wachholtz, 2013:26.
459 International Potato Centre CIP 2009; Potato Atlas.
460 Qu, Dongyu and Kaiyun Xie, eds. *How the Chinese eat potatoes*. Agriculture and food vol. 1. Hackensack, NJ: World Scientific, 2008: 23–24. http://site.ebrary.com/lib/alltitles/docDetail.action?docID=10688009.
461 Ibid.:23–24.
462 Ibid.:391
463 The Yangtze River floods of 1998 are responsible for the death of 4000 people and millions of people losing their homes. For more information on China floods, the author refers to the works of:
Ye, Qian, and MichaelH. Glantz. "The 1998 Yangtze Floods: The Use of Short-Term Forecasts in the Context of Seasonal to Interannual Water Resource Management: Mitigation and Adaptation Strategies for Global Change." *Mitig Adapt Strat Glob Change* 10, no. 1 (2005): 159–82.

potential and high nutritional value. Given the assumption that two metric tons of planted tubers can produce 40 metric tons of harvest per ha, the potato's value is above any grain crop in the world. Another reason for the potato's success is its long storage life and the fact that it can easily be transported over longer distances and longer periods of time.[464]

Some scientists believe that the early potato cultivation changed the history of Europe and was responsible for the rise of the West.[465] With the potato's ability to satisfy the demand of large-scale food supply, the potato contributed to 25 to 26 percent of the population growth between 1700 and 1900 in the Old World (Europe, Africa and Asia).[466] Soon after its introduction in China, the potato's value was also recognized. Together with foreigners and missionaries living in China, who cultivated the potato as a traditional European food, the expansion of the potato only continued. Attempts to increase the potato yield per mu were made as early as 1914 by the Department of Agriculture and Commerce of the National Government of China.[467] First experiments demonstrated that new varieties succeeded in higher yields, whereas older local tuber varieties planted in previous years showed stagnation in production.[468] The government initiated first breeding programmes in the early 1920s and the farmers in China were persuaded by the government to cultivate both the potato and sweet potato in order to strengthen the farming sector.[469] Real efforts to improve the varieties of potatoes in China were made in the 1930s when Chinese scientists imported new varieties of potatoes from America and Britain.

The Potato as a Staple Crop

As food security aspects of root and tuber crops date back over centuries, the advantages of the potato in terms of nutrition value have always played an important role in times of food crisis. In northeast China, during the Japanese invasion (1937–1945) the planting of sweet potato and potato for military

---

J.E. O'Connor and J.E. Costa. "The World's Largest Floods, Past and Present." *Science for a changing world* Circular 1254 (2004). Accessed October 18, 2015.
464   Qu, Dongyu and Kaiyun Xie, "How Chinese eat potatoes", 375.
465   Stamp, „…und weiss wie Alabaster", 27.
466   Nunn, N., and N. Qian. "The Potato's Contribution to Population and Urbanization: Evidence from a Historical Experiment." *The Quarterly Journal of Economics* 126, no. 2 (2011): 593–650. Accessed October 4, 2015. doi:10.1093/qje/qjr009.
467   Ibid.:373
468   Ibid.:373
469   Qu, Dongyu and Kaiyun Xie, "How Chinese eat potatoes", 373.

purposes was enforced.[470] This helped to increase the cultivation area and also introduced new varieties like *Okinawa 100* and *Norin 4* in 1941.[471] More heavily sown areas and an increase in tuber yield could be observed during the famines of 1959 to 1961.[472] During the devastating decades of Mao's political chaos, the "Grain First" campaign was the answer to the Great Chinese famine, induced by the "Great Leap Forward", when China transformed nature to serve human needs. Placing this focus on grain, the national system for storage was marked by poor management, which led to rot. This resulted in large-scale suffering due to this massive food shortage and millions died. The Chinese attempted to fight this food shortage through the planting of tubers in their small, home gardens. Over the 1960s, the sweet potato would replace corn as the crop of the poor.[473] The fact that tubers grow underground, helped the population to defence themselves from governmental assaults and confiscation of crops like corn, grain and rice, which were easy to discover by state agents.[474] The sudden expansion of potatoes was mainly driven by natural disasters like droughts and floods, which in turn resulted in severe food shortages. Often potatoes were planted after the failure of other crops.[475] With the development and cultivation of grain as a staple crop Yi Wang explains, "The role of potato had changed, and was becoming less important as a main food source in general, but it remained a major energy source for the very poor people who are malnourished or undernourished. This is particularly true for those, who live in the impoverished mountain regions"[476] of north China.

With the first Five-Year Plan after the foundation of the People's Republic, the government decided to support the cultivation of tuber plants and the increase of areas being harvested.[477] By 1960, already 30 institutes were actively working on China's potato research. In the 1980s attempts at creating

---

470 Gitomer, Charles S. *Potato and sweetpotato in China: Systems, constraints, and potential.* Lima, Peru: International Potato Center Regional Office [u.a.], 1996:24.
471 Ibid.:80; 81.
472 Gitomer, "Potato and sweet potato in China: Systems, constraints and potential", 11.
473 Thaxton, Ralph. *Catastrophe and contention in rural China: Mao's Great Leap forward famine and the origins of righteous resistance in Da Fo Village.* Cambridge, New York: Cambridge University Press, 2008:224.
474 Ibid.:224
475 Perkins, "Agricultural Development in China", 48.
476 Yi Wang "*Overview of potato production in China*". CIP-China Liaison Office in Beijing (http://www.cipotato.org/MF-ESEAP/Fl-Library/Pto-China.pdf).
477 Ibid.:376

genetically modified varieties were being made and research scientists began exchange information with foreign specialists and were working on international cooperation. In 1985, the International Potato Center in Beijing was established as a sub-division of the *Centro Internacional de Papa*, situated in Peru. After the introduction of Deng Xiaoping's Reform and Opening Up policies (改革开放), a more western lifestyle and accompanying eating habits started spreading across China and the market opened to western-style fast food. The demand rose in turn and frozen French fries (FFF), chips, and other potato-based food became increasingly popular among young Chinese.[478] According to Yi Wang, for many years China "remained as a net importer of frozen French fries from several sources in the international market."[479] But the growing demand for western-style food and dishes forced the Chinese potato industry to grow rapidly and offer domestic products for the new dietary habits. With the changes in food production and consumption, the use and importance of the potato changed as well. The potato remained significant in several different aspects: for food security, industrial purposes like the production of alcohol and animal feed.[480]

From 1961 until now, potato production including tuber-seed production has accelerated continuously.[481] Today, China is the world's largest potato-producing nation in the world (see Fig. 5). Taking into consideration that 60 percent of China's arable land is dry and lies in regions with low or less precipitation, where none of the common grain crops can grow without immense additional irrigation effort, planting drought and water-stress-resistant potatoes has been and remains the only solution to establish a domestic tuber production.[482] Hence potato, buckwheat, millet, and oats are the major crops and are cultivated either in inter or rotation cropping systems. Nevertheless the potato has the highest yield potential, is the most drought resistant plant and with the most efficient system for photosynthesis.

---

478 Ibid.:378
479 Ibid.:13
480 *Sustainable livelihood for rural households: Contributions from rootcrop agriculture UPWARD annual review and planning workshop 19–22 October 1997, Hanoi, Vietnam*. Makati City: Users' Perspectives With Agricultural Research and Development, 1998:63.
481 Data of China's potato yield and production are available since 1961.
482 Ibid.:394

Fig. 5: Potato Producing Countries in the World 2012

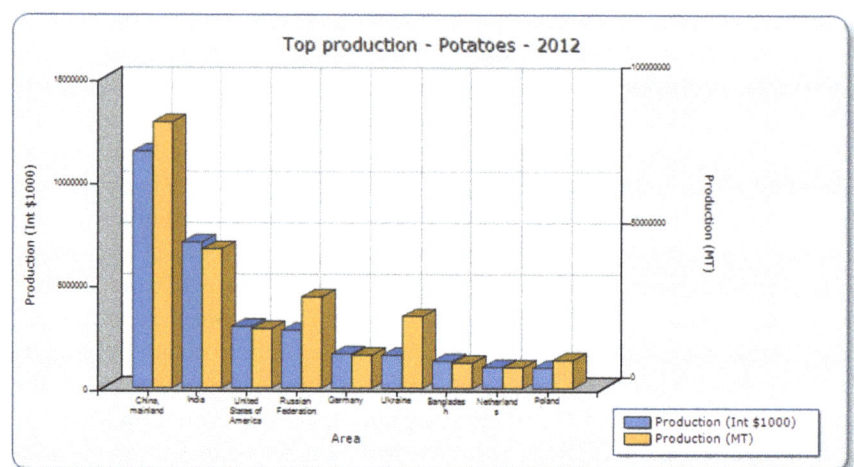

Source: Food and Agriculture Organization of the United Nations, http://faostat.fao.org/site/339/default.aspx

Potato Use

Traditionally the northern nomadic tribes in China mainly consumed meat (mutton, beef and goat), milk and other dairy products provided by their livestock. Nevertheless, throughout history interethnic trade relations ensured that the Mongolian diet included carbohydrates and vegetables as supplements. The Han-Chinese peasant society consumed cereals and vegetables, rice and grain as staple and rarely ate meat. Over history, the primary distinction between the North and the South diminished, but it remains true, that flour products are still more common in the North and rice dominates in the South.

In China the potato is important as a staple food and cash crop for millions of people. In particular, it appears that the contribution of potato production to household incomes seems to correlate with altitude and poverty. Meaning that in areas with high altitudes like the northern provinces of China, Inner Mongolia, and Shanxi, potatoes are cultivated as cash crops and staple food in order to secure a financial living for the local population. In rural areas, where jobs in industry or the service sector are rare, and where people typically live from agriculture or livestock, the cultivation of potatoes is essential. Especially for people living in the mountains the potato is the major source of income and sold at local markets or to brokers.⁴⁸³ In contrast, in lower altitudes tubers are often cultivated in small-scale farming and only for self-consumption as a staple food.

Processed potatoes, like potato starch is, according to the USDA Foreign Agricultural Service, "supported by the food processing sector and other industrial sectors such as textiles, paper milling, chemical, and pharmaceutical products."⁴⁸⁴ The production of high quality potato starch covers a major sector of the potato processing industry in China. Potato starch and flakes (for mashed potatoes) are popular products for Chinese consumers and this processing adds value to the potatoes because they can be used as raw materials for several other products including in "the development of food, papermaking, and textile, pharmaceutical and chemical industries."⁴⁸⁵ This contemporary form of potato processing and the accompanying industry began in the 1980s, with the import of more than twenty production lines from Europe and America. Since 2007, China has established several new potato processing plants, which produce "refined potato starch, potato chips, French fries and potato flakes."⁴⁸⁶ As Xie Kaiyun explains, the "potato is one of the seven major crops in China" and ranks fourth behind rice, maize, and wheat in regard of dedicated acreage for growing.⁴⁸⁷ With the growth of the total production since 2007, the potato became an integral part in the Chinese diet, but compared to other countries, China's potato consumption is still lower. With the increasing globalization of the last few decades, new and foreign processed foods were able to conquer China. Where previously eating potatoes was, according to Dwight H Perkins, "considered to be an act of desperation preferable only to starvation,"⁴⁸⁸ especially in the urban and suburban areas of metropolises like Shanghai or Beijing, where many foreigners also tend to reside, consumption habits have changed completely from earlier decades. Though the eating habits of urban dwellers under the influence of both foreign

---

483 "Sustainable Livelihood for Rural Households: Contributions from Root crop Agriculture", 63.
484 Ryan Scott, and Zhang Lei. "China – Peoples Republic of: Potatoes and Potato Products Annual." GAIN Report 12076 Unpublished manuscript, last modified January 18, 2016. http://gain.fas.usda.gov/Recent%20GAIN%20Publications/Potatoes%20and%20Potato%20Products%20Annual_Beijing_China%20-%20Peoples%20Republic%20of_12-20-2012.pdf.
485 "Development plan for the Potato Processing Industry during the "Twelfth Five-Year Plan" Period".
486 Qu, Dongyu and Kaiyun Xie, "How Chinese eat potatoes", 400.
487 Qu, Dongyu and Kaiyun Xie, "How Chinese eat potatoes", 94.
488 Perkins, "Agricultural Development in China", 48.

food habits and government reforms are slow to change. While the potato is becoming more important, it remains largely a side dish and has yet to full take on the same importance as rice or wheat.[489] But in the remote rural areas, the consumption of potatoes as a staple food is still very high, comparable with other countries in the world, where grain crops like millet, tubers (manioc) or rice – as carbohydrates – for energy-supply, still compose most of the diet.[490] Indeed, one of the major reasons why the potato plays an essential role in poverty elimination and food security is because of its high concentration of dietary minerals and vitamins.[491]

A further sector using and processing tuber crops is the energy sector. After the Chinese government in 2007 banned the extraction and production of grain-based alcohol made from staple crops like grain, rice or maize, potato, sweet potato and cassava became energy crops. Therefore roots and tubers became important resources as an energy supplier via the production of alcohol.[492] The underlying thought of the Government to push this law was the huge population and limited arable land demanding for a sufficient and secure food supply by grain crops.[493] Alcohol extraction for the ethanol biofuel production from tubers like cassava and sweet potato are common, but also the relevance of potato-use in the energy sector continues to accelerate. Cassava is only suitable to plant in the hot southern tropical and sub-tropical regions of China, whereas

---

489 "The popularity of western-style food has created strong market demand for FFF and other potato based products. Fabricated potato chip makers are the major users of dehydrated potatoes; dehydrated potatoes are also used in other snack foods and in Western style dishes like mashed potatoes" USDA Foreign Agricultural Service (Report 2012).
490 Yi Wang. "Overview of potato production in China: CIP-China Liaison Office in Beijing." Accessed May 12, 2014. http://www.eseap.cipotato.org/MF-ESEAP/Fl-Library/Pto-China.pdf.
491 Nunn, N., and N. Qian. "The Potato's Contribution to Population and Urbanization: Evidence From A Historical Experiment." *The Quarterly Journal of Economics* 126, no. 2 (2011): 593–650, p. 594. Accessed October 4, 2015. doi:10.1093/qje/qjr009.
492 Zhang, Heling, and Bofu Song. *Seed Potato Production in China*. 300th ed. Huhehaote, Inner Mongolia: Inner Mongolia University Press, 1992.
493 Qu, Dongyu and Kaiyun Xie, "How Chinese eat potatoes", 395.

the sweet potato grows in most of the warm regions.[494] But in the northern cold regions of China, it is the potato, which is best adapted. In fact, Tibet is ranked in first place in terms of yield and planting area,[495] Though the potato planting area is the smallest of all provinces (ranked 22nd) with just 600 ha, the yield per ha is 58 tons, reaching European and US averages, indicating that the Tibetan production is highly productive, compared to the Inner Mongolian average of 14.9 metric tons per ha in 2006. The potatoes use for alcohol production in Tibet (4.5 to 6 tons/ha) might be the result of the difficult transportation in the high plateaus of Tibet and the poor storage capacities and therefore for own consumption and not for export. Indeed, the potato is largely produced for alcohol extraction rather than as a staple food supply.[496] Therefore, the production of potato-based alcohol is an additional income for the farmers and increases the value of the potato.

---

494   For further information on using tubers for ethanol production, see:
Li, Shi-Zhong, and Catherine Chan-Halbrendt. "Ethanol production in (the) People's Republic of China: Potential and technologies." *Applied Energy* 86 (2009): S162-S169. Accessed January 18, 2016. doi:10.1016/j.apenergy.2009.04.047.
495   Qu, Dongyu and Kaiyun Xie, "How Chinese eat potatoes", 396.
496   Ibid.:396

Tab. 9: Potato Planting Area, Production and Yield 2006

Region	Area (ha)	Rank	Production (t)	Rank	Yield (t/ha)	Rank
Guizhou	592800	1	7730000	5	13.04	16
Inner Mongolia	589100	2	8795000	12	14,93	13
Gansu	567800	3	9400000	1	16,56	11
Yunnan	539900	4	8610000	3	15,95	12
Sichuan	348200	5	8515000	4	24,45	4
Chongqing	347400	6	4706000	6	13.54	15
Heilongjiang	319300	7	4050000	7	12,68	18
Shanxi	299200	8	2415000	10	8,07	22
Shaanxi	252000	9	2550000	9	10,12	20
Hubei	229800	10	3390000	8	14,75	14
Ningxia	186600	11	1620000	15	8,68	21
Hebei	152900	12	1890000	12	12,30	19
Jilin	1421900	13	1850000	13	13,04	17
Hunan	113200	14	2085000	11	18,42	9
Liaoning	90000	15	1685000	14	18,72	8
Fujian	87400	16	1510000	17	17,28	10
Qinghai	80100	17	1545000	16	19,29	7
Guangdong	42800	18	940000	18	21,96	6
Xinqiang	22900	19	755000	19	32,97	2
Anhui	7600	20	185000	20	24,34	5
Jiangxi	4200	21	105000	21	25,00	3
Tibet	600	22	35000	22	58,33	1
TOTAL/AVERAGE	**5015700**		**74355000**		**14,82**	

Source: Qu, Dongyu and Kaiyun Xie, eds. *How the Chinese eat potatoes*. Agriculture and food vol. 1. Hackensack, New Jork: World Scientific, 2008.

Development of the Potato Production in China

Looking at the development of other cash crops in the last decade, the areas for planting rice and wheat have decreased, whereas maize and potato show an increase in the planting area and production. Due to the circumstance that the yields of the three major cash crops are exhausted and exceeded because both arable and fertile land is limited, potatoes offer a new solution due to their typically high production and yield rates.[497] However, in the last ten years potatoes yield have not increased in relation to the production and the area harvested. This should be seen as "China's Potato Problem". The table below shows that potato production in metric tons and the area harvested increased up to four times from 1961 to 2007. It is clear, however that the yield stabilized at low levels, and did not significantly rise relative to production. For example, 70 million metric tons were produced in 2007 on 5 million hectares with an average of 15 metric tons per hectare. Despite the fact that the area harvested tripled and the production facilities have more than quadrupled, the amount of yield shows a stagnation existing since 1992. In the past, there was only a minor increase in yields of 5 million metric tons since 1973, whereas the production of potatoes in China has continuously increased.[498] This discrepancy highlight China's yield gap.

Low productivity since the early 1990s has been a severe problem for China. Although China increased the acreage for potato production, the yield remained unchanged at low levels. China's potato yield only reaches 30 to 40 percent of the US and European average.

Tab. 10: *Potato Production in China 1961–2009*

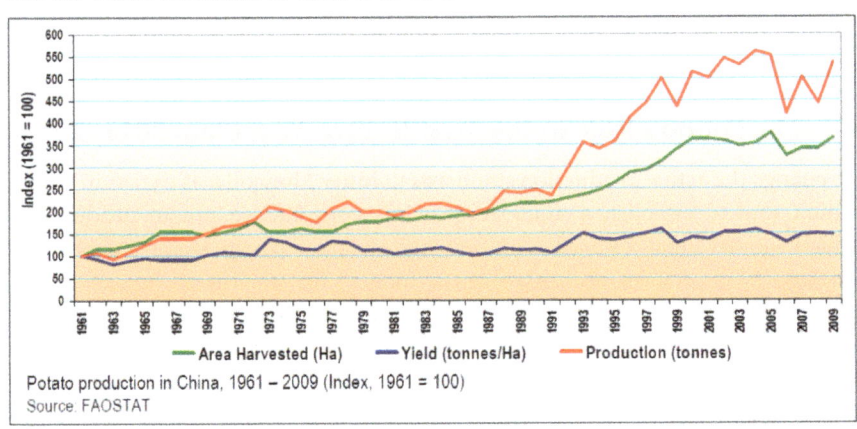

Source: CIP Annual Report (online www.internationalpotatocenter.com)

---

497 Qu, Dongyu and Kaiyun Xie, "How Chinese eat potatoes", 392.
498 Data provided by FAQ and China's Agricultural Yearbook.

China's potato industry is based on the production of small holders, who in some cases cultivate less than 1ha (15mu亩) of land without the use of modern agricultural equipment. On the production line, many processes are typically completed by hand, from sowing to harvesting and packaging. This makes market access for the smallholding farmers even more difficult because there is no quality control or production consolidation, which makes the potatoes competitive on the market. Furthermore, China lacks an established seed sector similar to other emerging markets, like Sub-Saharan Africa and Latin America. Instead, farmers keep seeds and tubers from previous harvests or buy uncertified seed from multiple poorly qualified sources.[499]

China's potato production is inefficient in means of yield and cultivated acreage. This problem stems from different issues regarding the whole production line, beginning with unqualified seeds, un-sustainable cultivation and storage, as well as inefficient post-harvest management. Lack of agricultural knowledge, false disease treatment and incorrect tuber seed management are also responsible for the disparity between potato yields in China and those in Europe. Though Germany's harvested area is markedly smaller than that in China, the relative yield to the production area is higher. Intelligent crop solutions beginning with sustainable and healthy seed production, effective cultivation, and an increase in yield are a major challenge China must face in the coming years in order to meet the requirements of a sustainable potato production.

6.1 The Potato's Role in the Twelfth Five-Year Plan, 2011–2015

To change the actual production standards in Inner Mongolia, as well as in whole China, and to establish a modern, growing, consolidated potato production, China's government released a five year strategic plan to help to alleviate poverty, improve food security, and increase the income for small-farm households.[500] The Twelfth Five-Year Plan proposes changes that should lead to the reduction of people living in poverty by half. The potato is officially one of the crops utilized in this ambitious plan. According to the Twelfth Five-Year Plan, the potato processing industry will step into a new stage of development. The "Development Plan for the Potato Processing Industry during the Twelfth Five-Year Plan

---

499   Confidential information from the Interview with the Head of the Agricultural Bureau in Hohhot, see appendix vi.
500   For details, see full document in Chinese at: http://www.moa.gov.cn/ztzl/shierwu/hyfz/201112/t20111227_2444181.htm.

Period" are obligatory for the period of 2011–2015. The outline of the political guideline incorporates the structural adjustment and development of China's potato processing industry in the next five years, with the aim to realize the sustainable development of this industry.[501]

The Chinese Government therefore chose to pursue the expansion and increase of the potato production to ensure food security and to eliminate poverty.[502] In 2010, the number of potato processing enterprises increased from ten in 2005 up to twenty-five.[503] The output of the top five manufacturers of potato starch shared 15 percent of the total yield in 2005, increasing to 23 percent of the total yield in 2010.[504] During the Eleventh- Five-Year Plan first attempts to promote the national potato production were already made.[505] The potato processing industry in China accelerated its pace of structural adjustment and technological progress, as well as the implementation of a series of major scientific and technological projects (*Small Farmers Adapting to Global Markets* SFAGM, supported by the World Health Organization WHO) closely related to the potato processing industry.[506] Quality standards for food safety and production have continuously increased

---

501 Confidential information, "Development plan for the Potato Processing Industry during the "Twelfth Five-Year Plan" Period".
502 马铃薯加工业"十二五"发展规划. 5:"维护粮食安全为马铃薯加工业创造了良好的发展空间.粮食安全问题是关系经济安全和国计民生的重大问题,《国家粮食安全中长期发展规划纲要（2008-2020年）》明确将马铃薯作为保障粮食安全的重点作物,摆在关系国民经济和"三农"稳定发展的重要地位". Document available online at: http://www.chinagrain.org/userfiles/23_20131009023.pdf.
503 马铃薯加工业"十二五"发展规划, 2:"变性淀粉、全粉、薯片加工企业数量显著增加,规模以上企业均由 2005 年的 10 家左右上升到 2010 年的 25 家以上".
504 Confidential information, "Development plan for the Potato Processing Industry during the "Twelfth Five-Year Plan" Period", 4.
505 马铃薯加工业"十二五"发展规划, 3: "十一五"期间,马铃薯加工业的产品结构进一步优化. 2010 年,马铃薯加工业消耗马铃薯 692 万吨,比 2005 年提高 28.4%,年均增长 5.1%. 以品种计,马铃薯加工产品的产量为：淀粉 45 万吨,变性淀粉 16 万吨,全粉 5 万吨,冷冻薯条 11 万吨,各类薯片 30 万吨,粉丝、粉条、粉皮 30 万吨（以干基计）".
506 马铃薯加工业"十二五"发展规划, 4: "... [...]...期间, 我国马铃薯加工业结构调整、技术进步步伐加快,一批具有较强竞争实力和技术优势的骨干企业发展壮大... [...]...组织实施一批与马铃薯加工密切相关的食品加工重大科技专项,在食品用马铃薯蒸汽去皮及水力切条、提高马铃薯淀粉提取率和节水率、变性淀粉生产应用技术开发等关键技术领域取得了突破.

and have become an inherent part of the "Food Safety Law of the People's Republic of China" since 2009.[507]

In 2010 the consumption of processed potatoes increased by 28.4 percent. The output of processed potatoes has varied with these newer diet habits, showing enormous growth. The *Development plan for the Potato Processing Industry during the "Twelfth Five-Year Plan" Period* reported that in 2010 "450,000 tons of starch, 160,000 tons of modified starch, 50,000 tons of potato powder, 110,000 tons of frozen French fries, 300,000 tons of various potato chips and 300,000 tons of bean starch noodles/sheets and vermicelli (on dry basis)" were produced by Chinese companies.[508]

To fulfill this ambitious plan to develop the potato industry, China had to grapple with severe problems regarding production processes due to problems with unstable raw material resources. Even though China ranks number one in worldwide production of potato products, the relative production quality compared to non-Chinese markets is quite low.[509] Multiple problems plague the potato seed tuber market in China. First, the proportion of virus-free seed potatoes and the unit-yield level is relatively low. Second, the stable supply of raw materials for potato processing faces severe challenges and the working conditions are poor. Third, harvesting methods, storage and transportation equipment are relatively lagging behind in terms of technological developments. Fourth, the losses caused by disease and rot are rather serious and the special seed potato for processing has been insufficient. Further, poor quality is common among the raw materials, which ultimately leads to a poorer end product. The extreme contradictions between the "small-scale farming" of potatoes characterized by small-scale planting and decentralized management, and "large-scale production" characterized by mass production and market-based operations are major issues to be solved in the coming years. The large-scale and mechanized planting pattern utilized in mass production methods is still deficient and the construction of industrial bases remains weak.[510]

---

507 马铃薯加工业"十二五"发展规划, 5: "随着全社会对食品质量安全的日益重视和《中华人民共和国食品安全法》及其实施条例的颁布实施,"十一五"期间, 马铃薯加工业食品安全水平和产品质量不断提高".
508 马铃薯加工业"十二五"发展规划, 3: "2010 年, 马铃薯加工业消耗 [...] 提高 28.4%".
509 马铃薯加工业"十二五"发展规划, 7: "原料保障压力增大。我国马铃薯总产量虽位居世界首位, 但发展水平有待提高…[...]".
510 Confidential information (Syngenta), "Development plan for the Potato Processing Industry during the "Twelfth Five-Year Plan" Period", 6.

The industrial structure is underdeveloped and disadvantageous. Although the structure of potato processing industry in China has been improved since the Eleventh Five-Year Plan 2006–2010, but it is still not very effective in several aspects.[511] The proportion of primary processing products is relatively large while the variety and output of high value-added products is still small and in need of expansion. The number of enterprises with advanced technical equipment and mass production is limited. At the same time outdated industrial capacities with simple and crude equipment, outdated technology and low quality and production rates still occupy a large proportion of the market. Governments and Industries Research and Development (R&D) on the integrated utilization of by-products is still being developed and the industrial chain of a circular economy[512] has not yet formed, which limits the healthy development of potato processing industry.[513]

Society's demand for sustainable development has risen according to increases in wealth. However, China is lagging in regards to environmental standards, while the demand for sustainable development is increasing. In view of the requirements for the construction an energy saving and environmentally friendly society as well as sustainable resource management and environmental constraints, is now an urgent need to realize these advances. The implementation of energy conservation, emission reduction, and integrated resource utilization in the potato processing industry are of high priority. This can be achieved by pushing the transformation of development through the utilization of new technologies, new processes, and new equipment.[514]

Key tasks for the implementation of the Twelfth Five-Year Plan were the guarantee of raw material supplies and optimizing the regional distribution of potato starch and powder processing industry in the main producing areas.[515] Therefore it was necessary to improve the quality of tuber seeds and promote different varieties

---

511　Ibid.:1
512　A circular economy refers to a sustainable economy model, which is the opposite of the linear model (*take, make, dispose*) and includes the use of renewable resources and energy, recycling of goods and waste as well as the reduction of pollutants to enter the biosphere.
513　Ibid.:7
514　马铃薯加工业"十二五"发展规划, 7: "可持续发展要求提高. 按照建设资源节约、环境友好型社会的要求，资源环境约束日益强化，运用新技术、新工艺、新材料、新装备推动马铃薯加工业实现节能减排、资源综合利用的要求不断提高，加快转变发展方式尤为迫切.
515　马铃薯加工业"十二五"发展规划, 12:" 强化原料供应保障"…"优化行业区域布局，在马铃薯主产区及其周边地区，发展马铃薯淀粉和全粉加工. 在大中城市及其周边地区，发展薯片、薯条、变性淀粉及其他高附加值产品".

of virus-free potatoes.⁵¹⁶ Support of the R&D sector was targeted, according to the "Development plan for the Potato Processing Industry during the "Twelfth Five-Year Plan" Period", in order to "enhance the cultivation of scientific research talents and construction of R&D teams."⁵¹⁷ The cooperation and exchanges between enterprise, universities and research institutions were a further key step to "improve [the] technological achievements in order to promote the independent innovation capability of China's potato processing industry in a comprehensive way."⁵¹⁸

Implementations by the Chinese government

To reach the goals of the Twelfth Five-Year plan China still implements a sustainable crop management program in different cropping models and varied water-saving irrigation methods are supported.⁵¹⁹ The water-saving drip irrigation system supersedes the widespread pivot-irrigation, which is now banned and no longer supported by the Chinese Government.⁵²⁰ Further sustainable cultivation technologies, like crop rotation, minimizing pesticides and fertilizers are also implemented. With the establishment and development of a sustainable potato production technology, China is forcing businesses to meet the requirements of national standards.⁵²¹ The reduction of emissions, pollution control and energy saving methods are currently of the highest priority.⁵²² The government further encourages the "No/Less tillage culture" method, which means that crop residues are used to protect the new generation and the soil cover doesn't get destroyed by tillage, in order to prevent soil erosion and to support a sensitive application

---

516 马铃薯加工业"十二五"发展规划, 12:" 促进马铃薯种植业的种薯脱毒化、品种专业化、种植规模化和机械化，保障加工专用薯的种植面积及单产的稳步增长.加大对马铃薯原种生产补贴、规模化集约经营和高产创建的支持力度，推广应用脱毒种薯和配套高产栽培技术".

517 Confidential information (Syngenta), "Development plan for the Potato Processing Industry during the "Twelfth Five-Year Plan" Period", 11.

518 Confidential information (Syngenta), "Development plan for the Potato Processing Industry during the "Twelfth Five-Year Plan" Period", 11.

519 Confidential information from the Interview with the Head of the Agricultural Bureau in Hohhot, see appendix vi.

520 Ibid.:vi.

521 马铃薯加工业"十二五"发展规划, 15:" 加快马铃薯清洁生产技术的开发和推广，确保污染物排放和节能降耗达到国家相关标准要求".

522 马铃薯加工业"十二五"发展规划, 15:" 加大马铃薯加工行业节能降耗、减排治污工作的推进力度，鼓励马铃薯加工企业建立与加工规模相适应的污水处理和综合15利用基础设施".

of fertilizers to prevent soil acidification.[523] Developing the farming machineries for small famers and modernizing the post-harvest potato handling systems to consolidate the production is a further measure to help farmers to gain access to local markets.[524]

The acceleration of financial support for the potato processing industry is also described in the Twelfth Five-Year Plan.[525] According to insider sources, the financial aspect of the "Development plan for the Potato Processing Industry during the "Twelfth Five-Year Plan" Period" plans to "expand the subsidies to potato original seed production and the scale of high yield trial demonstration and support the development of potato planting industry to increase potato output and ensure raw material supply."[526] The proposed plan is to take full financial support from the central and local government for major projects in the potato processing industry, such as the implementation of public service platforms for farmers, guaranteeing the technological transfer and know-how within the involved enterprises, supporting strategies for energy-saving and emission reduction as well as the transformation to a clean and independent sustainable production.[527] The Agricultural Development Bank of China and commercial banks will provide credit support to potato processing projects and enterprises for technological transformations, which meet with national industrial policies and conditions for loans.[528] To reduce the amount of unproductive farmland, the government further supports land aggregation. Two policies for land consolidation have been discussed: higher subsidies for families or farms working together and subsidies for certified, virus-free tubers.[529] The fragmentation of cultivated land often results in problems of land loss, restrictions

---

523 Confidential information (Syngenta), "Development plan for the Potato Processing Industry during the "Twelfth Five-Year Plan" Period", 12.
524 Ibid.:16
525 Ibid.:15
526 Ibid.:15
527 马铃薯加工业"十二五"发展规划, 17:"加大中央和地方财政支持力度，充分利用现有财政政策及资金渠道，对马铃薯加工产业和产业集群公共服务平台建设、企业技术改造、节能减排、清洁生产、重点装备自主化和自主品牌建设等重点项目给予支持".
528 马铃薯加工业"十二五"发展规划, 17:"中国农业发展银行及商业银行对符合国家产业政策和贷款条件的马铃薯加工项目和企业技术改造提供信贷支持，对实力强、资信好、效益佳的企业优先安排贷款，增加授信额度；支持符合条件的马铃薯工业企业通过在银行间债券市场发行短期融资券、中期票据、中小企业集合票据等方式拓宽融资渠道，募集生产经营资金".
529 Confidential information from the Interview with the Head of the Agricultural Bureau in Hohhot, see appendix vi.

to irrigation access, difficulties in the application of pest control and poor land-use management. As a consequence, environmental damages like soil erosion, soil-quality degradation and water pollution increase. The environmental impact of cultivated land degradation will threaten the sustainability of food production. Through the creation and enforcement of policies for land consolidation and land productivity, soil quality and irrigation systems can be improved.

The implementation of the Twelfth Five-Year Plan will be under the charge of the Ministry of Industry and Information Technology together with the Ministry of Agriculture. International collaboration with seed producers and research and development enterprises are of utmost interest for the development of a sustainable, competitively viable potato market. This market must be capable of coping with the ever-increasing food demand and international quality standards as well as helping to alleviate poverty and secure the social and economic well-being of the Chinese farmers.

For Inner Mongolia the implementation of the Twelfth Five-Year Plan is of high interest because the potato is the dominant crop in Inner Mongolia, where more than 10 percent of China's total potato production takes place. With the support of the government and international experts the goal is to double the potato production by 2015 in order to help to further develop the region socially and economically. Inner Mongolia is still one of the poorest provinces of China with areas that are lagging behind in terms of infrastructure, social and economic well-being, and access to international markets. Although Inner Mongolia's economy is dominated by heavy industries, like coal mining, rare-earth extraction and chemicals as well as meat and dairy production, the population does not benefit from these economic industries as the majority lives from own subsistence in remote and rural parts. Therefore developing the agricultural sector, which is still the main income source for most of the families in Inner Mongolia is one of the major goals of the Twelfth Five-Year Plan.

6.2 The Potato's Role in Inner Mongolia – More than Just a Staple Food

China produces significant amounts of both of these tubers. In 1991, 12 percent of the world's share of potatoes were cultivated in China, and 80 percent of the world's sweet potatoes, which indicates China's prominent role in the tuber crop market in the last twenty years.[530] The trend line for root and tuber production as

---

530 Gitomer, "Potato and sweet potato in China: Systems, constraints and potential", 3.

well as for tuber yield, as Gitomer described in 1994, has continuously increased, until today.[531] Potatoes as well as sweet potatoes contributed to China's national grain output as Perkins pointed out, because of their high yields and adaptability to different soils.[532] Especially in terms of calories per mu, "the potato is roughly twice [as] productive as other dry-land crops" and therefore an ideal alternative to other cash-crops in the Northern provinces of China.[533]

China's potato production can be distributed into four different "agro-ecological" zones, as the following table shows.[534]

Map 6: China's, Agro-Ecological Zones of Potato Cultivation

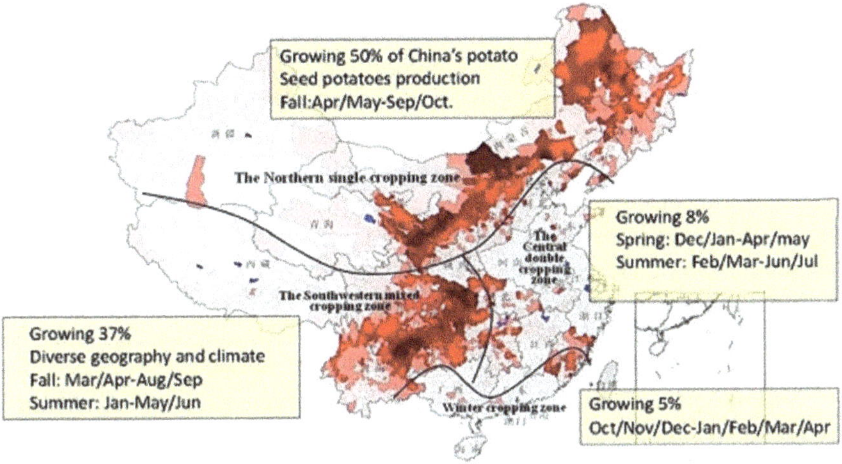

Growing 50% of China's potato Seed potatoes production Fall:Apr/May-Sep/Oct.

The Northern single cropping zone

The Central double cropping zone

The Southwestern mixed cropping zone

Winter cropping zone

Growing 8% Spring: Dec/Jan-Apr/may Summer: Feb/Mar-Jun/Jul

Growing 37% Diverse geography and climate Fall: Mar/Apr-Aug/Sep Summer: Jan-May/Jun

Growing 5% Oct/Nov/Dec-Jan/Feb/Mar/Apr

Source: Overview of potato production in China. Potatoes are widely distributed throughout the country in four main agro-ecological zones (Source: YAAS, 2015). Available at: http://www.potatopro.com/china/potato-statistics

Heilongjiang and Inner Mongolia developed into major potato seed stock producers in the last decades because of the developed infrastructure and their good quality seed tubers.[535] The regions is known as the "Northern Single Cropping Zone," where more than 40 percent of the potato crop is produced for fresh

---

531 Ibid.:3
532 Perkins, "Agricultural Development in China", 49.
533 Ibid.:49
534 Gitomer, "Potato and sweet potato in China: Systems, constraints and potential", 104.
535 Ibid.:94

consumption and processing.[536] It is the largest of the four production areas, including northern Hebei, Liaoning, Jilin, Heilongjiang, Ningxia Huizu, Gansu, Qinghai, and Xinjiang Uighur. As mentioned previously, the harsh climate makes this region perfectly suitable for potato cultivation. The irrigated parts of Inner Mongolia's plateau have become an especially important player in the production of processing quality potatoes, with an upward trend predicted for the coming years.[537]

Usually only one crop is planted in April or May and harvested between September to October. This system is dependent on the short growing season of ninety to 130 days with overall relatively low temperatures. The recommended practice is to rotate the crops in the following year between potato/grain, potato/rapeseed, or potato/sunflower. The general climate of the "Northern Single Cropping Zone" is marked by cool summers (24°C) and long, freezing winters (-28°C). Because of the heavy temperature drops in fall late potato varieties cannot be planted. In spring the temperature increases quickly, increasing, the risk of droughts which can cause severe damage to the tubers. Thus timely planting and harvesting in spring and autumn is essential. The tuber seeds are commonly stored for six months or longer before being replanted in spring. Therefore qualified tuber storage is essential and may affect the outcome of the next crop generation. Disease and pest treatment are important, because the tubers can pass along the infection to the new plants. The need of healthy tubers is the main issue for Inner Mongolian potato farmers to increase their yield.

To secure stable yields, increase production and decrease the risk of exhausted soils by monocultures, the government emphasizes on intercropping and rotation of potatoes, which dates back to the same successful cultivation system developed by farmers since the time of early crop cultivation.[538] The major intercropping or rotation systems with potato are in combination with other grain crops, cotton or vegetables. Commonly in Inner Mongolia the potato often rotates with wheat, corn, millet or sorghum. The crop rotation system of Inner Mongolia has changed only slightly over the last fifty years. While wheat, oat, and buckwheat still dominate the mountainous areas, millet, flax, soybean, oat, and wheat are intercropped in the plains. Currently, the rotation with oil-crops like rapeseed and sunflower is widespread. For more information see table Siziwang Qi Banner.

---

536 Ma Shuping 2012 "*Current Development and Prospect: Potato Production and Processing in China*" (held at the International Potato Congress, Edinburgh, 2012).
537 Papademetriou, Minas K. 2008 "Workshop to commemorate the International Year of the Potato – 2008", FAO.
538 Gitomer, "Potato and sweet potato in China: Systems, constraints and potential", 101.

6.3 Case Study 1: Potatoes in Siziwang Qi Banner

During the field research in Inner Mongolia in Siziwang Qi in October 2013, thirty local farmers were interviewed, mainly concentrating on potato production. Twenty-nine of the interview subjects identified as Han, who represent the majority of farmers in Siziwang, though Mongolians are widespread and also work as farmers. This case study concentrates on the potato production in Siziwang, illustrates the different cultivation methods and analyzes, whether the farmers perceive their business as profitable or not.

Standardized interviews were chosen in order to ensure that every person who was interviewed had the same framework of questions and so that the answers would be more easily compared. The questions were however designed to allow for non-structured answers.

1. Do the farmers support crop-rotation and inter-cropping with other grains?
2. What kind of crops do they plant and why?
3. What is their main key driver for cultivating potatoes?
4. What is their return on investment (ROI) and for what do they invest their net-income?

Nearly ninety percent of the farmers answered that they do perform crop rotation (*Table 11*). The following table shows that the plants intercropped with potatoes are mainly grain crops and oil-crops. The farmers then either plant the potato in combination with sunflower and rapeseed in the following season. During the winter season from September to April, the field lies fallow. Only 10 percent of the thirty interviewed farmers produce for their own subsistence (Table 12) and do not sell their crops to the local markets, whereas the remaining 90 percent fully support the planting of potatoes in combination with other cash-crops. Grain and oil crops are mainly rotated with potatoes in order to reduce the risk of insect infestation or other diseases.[539] The cultivation of grain and oil crops further increases the income of the farmers.

---

539 Vreughendil, Dick 2007: "Potato Biology and Biotechnology. Advances and Perspectives", 545.

Tab. 11: *Crop Rotation and Intercropping among Thirty Farmers in Siziwang Qi*

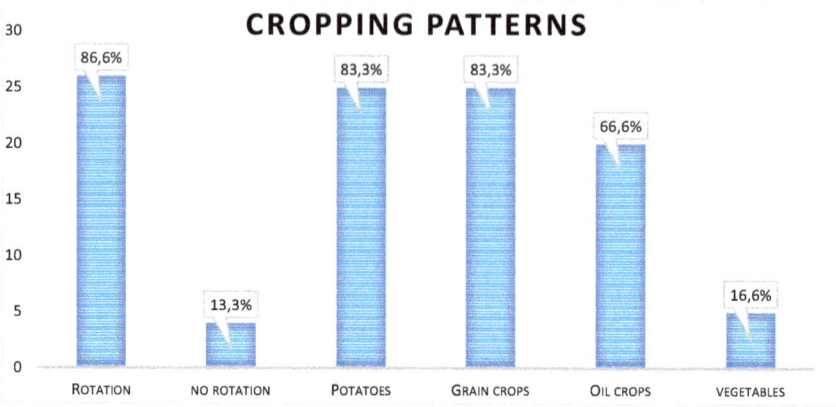

Source: Cilia Neumann 2016, (based on field research Inner Mongolia September 2013) "*What kind of crops are cultivated in Siziwang?*"

When asked the reason for potato cultivation the farmers explained that the potato is the most efficient crop to plant in Inner Mongolia and that the netprofit for potato production is higher and more profitable than for other crops (*Table 12*). Potatoes are planted because of their adaption to the Inner Mongolian environment. The farmers know about the advantages of the potato as a water-resistant and drought-resistant crop. The main reason for planting potatoes instead of other grain crops is that they are highly efficient (53 percent) and secure a higher income for the farmers in Siziwang (33 percent). Even the minority 10 percent, who plant potatoes for subsistence chose this crop because of its high nutrition.

Tab. 12: *Crop Selection among Thirty Farmers in Siziwang Qi*

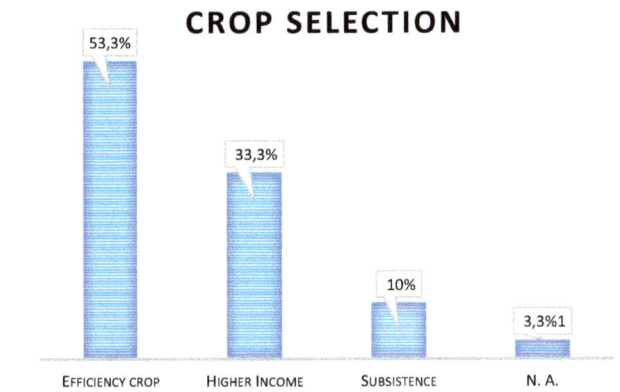

Source: Cilia Neumann 2016, (based on field research Inner Mongolia, September 2013) "*Why do the farmers in Siziwang cultivate potatoes?*"

Governmental subsidy might be another reason for planting potatoes and is a financial incentive for the rural population to invest in potato production. Ninety-three percent of the farmers reported that they are receiving governmental subsidies. It seems that re-investment for technical equipment, better irrigation systems, fertilizer and machines is very important for the farmers in Siziwang to improve their production standards, seed and tuber quality, and to increase their net-profit. But when it comes to the point of re-investment, approximately 30 percent make no re-investment in technological equipment, seed quality, or irrigation, whereas 46.6 percent re-invest mainly in machinery (harvest machines, tractors, seed machines), and less money is invested in irrigation technologies and seed care. Only 16.6 percent of the farmers re-invest in new dwell or drip-irrigation systems and in plastic mulch and greenhouses (*Table 13*). Little is re-invested in seed care and pesticides (6.6 percent), which astonishes, as an established seed market is a major contributor to improve yield productivity.

Tab. 13: Purpose of Re-Investment among Thirty Farmers in Siziwang Qi

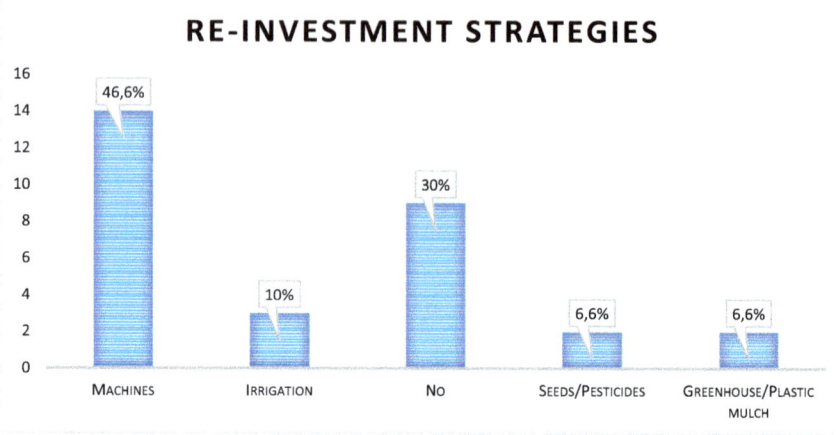

Source: Cilia Neumann 2016, (based on field research Inner Mongolia, September 2013) *"Usage of re-investment"*

The usage of heavy and huge harvest machines, tractors, and seed machines is only possible when the cultivated land is of a designated size or if the farmers consolidate their acreage. But most of informants had less than 100 mu (亩) acreage available to cultivate potatoes.

Tab. 14: Field Size among Thirty Farmers in Siziwang Qi

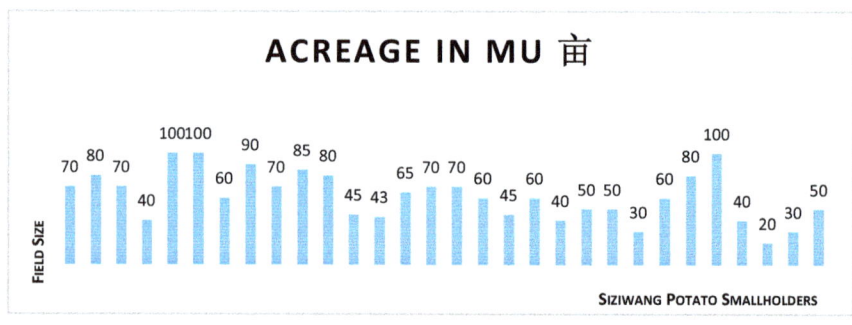

Source: Cilia Neumann 2016, (based on field research Inner Mongolia, September 2013) *"Acreage in mu 亩"*

The average size of potato fields in China is five acres (30.3 mu/2.2 ha). The technological equipment available for small-farm households is far behind Western standards. Often the farmers work with bare hands or animals in the fields, and using a tractor is not common.[540] The average size of acreage of the farmers interviewed in Siziwang Qi is 61.76 mu (4.12 ha/10.81 acres) – twice as much as country average. This reflects the strong development of potato production in Inner Mongolia, as well as the importance of the "Northern Single Cropping Zone," where the majority of potatoes area cultivated.

Only three farmers of the 30 interviewed are tilling 100 mu (6.6 ha) of arable land, but none of them manages more acreage than 100 mu. Therefore the investment in heavy and expensive machinery does not appear to be affordable for the other 50 percent, who stated that they make no investment in technological equipment. Instead, 23 percent of the farmers prefer to invest in irrigation projects, seed quality or in greenhouse/plastic mulch solutions, in order to improve the situation of chronic water-shortage, and to increase production and yield of the potato crops.

The situation found in Inner Mongolia as represented by the case study in Siziwang Qi mirrors China's larger problem of implementing a consolidated potato production. The status-quo shows that the potato production in Inner Mongolia is developing, but still has enormous problems in terms of homogenous production processes, sustainable cultivation methods and the use of modern technologies like seed and harvest machines, and thus becomes competitively viable on a global level.

Further, even though governmental policies at the national level provide guidelines for agricultural development, including mechanisms to stimulate the rural economy and supporting the agricultural production, the implementation of these measures are moving forward slowly. When possible the farmers try to progress their individual situation by taking small actions, like improving their irrigation method or investing in better seed quality and storage capacities. But without external support, in the form of international cooperation with specialists or governmental support the smallholding farmers remain unsuccessful and are poorly represented in the farming lobby. Therefore as mentioned before organizations like the WHO, United Nations Convention to Combat Desertification (UNCCD) and United States Agency for International Development (USAID) as well as the non-governmental Fair Labor Association (FLA) are supporting the local government and cooperate with international industries to improve the living and working conditions of smallholders in developing countries.

---

540 Crop ideas 26.4.2013, available online http://cropideas.blogspot.de/2013/04/chinese-potato-market.html "*Most of the work is done by hand, with animal use for traction or with small tractors*".

7. Inner Mongolia's Environmental Concerns – The Change of the Environment since 1950

The following chapter focuses on different forms of environmental degradation in Inner Mongolia. Through analyzing various development processes in Inner Mongolia society patterns emerge as to how Inner Mongolia's environment has changed since 1950. Economic demographics together with internal (Han-Mongolian culture) and external (different forms of land-use practices) factors have all influenced the local attitudes and values regarding environmental concerns. This chapter is divided into two parts. The first deals with specific and obvious environmental problems in Inner Mongolia and how these issues are being addressed. A short retrospective of the history of Inner Mongolia's environmental deterioration will be presented as well as the changes to Inner Mongolia's environment that have occurred and the general impact of land reclamation over the last sixty years. Two specific examples have been chosen to address these issues more closely. First, chapter 7.1 will introduce irrigation and water shortages as one of the major problems affecting farmers and herdsmen. Next, in chapter 7.1.1 the concept of *Water User Associations* (WUA) as a successful water-management tool based on user organization and decision making will be outlined. To visualize the problem of water-shortage insight into the irrigation problems of farmers in Siziwang will be provided in 7.1.2. The second example (7.2) to be presented refers to the problem of desertification and dryland expansion as a national issue connected to Inner Mongolia's environmental crisis due to the fact that dryland expansion leads to further decline in fertile land-resources. Here re-forestation, grassland restoration, and conservation methods on the basis of the latest *Grain for Green* program will be emphasized in order to illustrate the difficulties connected to implementing sustainable development in economically significant regions.

The relationship between societal factors and environmental damage will gradually be understood through Jiang Hong's analysis of culture and land relations, human driving forces, environmental awareness, and the societal responses of the people to environmental changes.[541] Societal factors have been shown to affect the environment and can be separated into specific activities such as cultivation, overgrazing, and deforestation but also into underlying driving forces including population changes, belief systems, economies, technology, and political

---

541 Jiang, "The Ordos Plateau", 40.

structures.[542] These varied driving forces and socio-economic factors can affect the environment either negatively or positively and therefore serve as responses to environmental degradation. Some of these factors mentioned in earlier chapters were responsible for large-scale environmental degradation in Inner Mongolia, especially between 1949 and 1978, when two political systems were propagated and followed by different reforms and campaigns under Mao Zedong and later in the reform era under Deng Xiaoping. With the changes in political organization after 1978, and the emphasis on economic development and new sensitivity towards environmental concerns, the same societal factors like traditional land-use practices can actually led to positive effects on the environment, but only as long as population and technological maintenance keep up with the development.[543] This increase in Inner Mongolia's economic growth also helped to alleviate poverty especially impacting the rural population. The political reforms of the HRS allocated the land-use rights to households, thus stimulating the individual's responsibility for economic self-sufficiency. The combination of the national policy of population control, new land-use regulations as well as economic and environmental development programs helped to reduce environmental degradation in Inner Mongolia. Though overall, Inner Mongolia and the rest of China have suffered from severe environmental damage, the net result of which since 1980 has led to initiatives for environmental improvement, ecological rehabilitation, and a rise in environmental awareness in the general public.[544]

Land reclamation for agricultural purpose had one of the greatest impacts on Inner Mongolia's environment due to the region's semi-arid climate and vegetation, which was largely responsible for the limited cultivation and made Inner Mongolia's agricultural development difficult. Farmers have confirmed, that without irrigation the cultivation of a fertile cropland there is impossible. As early as the nineteenth century, Han merchants settling in the Hetao Plain[545] (河套平原) have already realized that these harsh regions required an irrigation

---

542 Ibid.:64
543 Ibid.:171
544 Ibid.:172
545 Hetao Plain is a vast area in Northern China along the upper reaches of the Yellow River, including plains and plateaus on both sides of the Yellow River Loop. Often it is referred precisely to the western and eastern parts, called Xitao and Dongtao. Hetao Plain is bounded by the Yin Mountains in the north, the Great Wall in the south, the Alashan in the west, and the Lüliang in the east. It includes the Yinchuan Ningxia Plain, the Ordos Plateau, and the Loess Plateau and forms parts of Shaanxi, Ningxia, and Inner Mongolia.

system to provide the fields with enough water during the dry seasons[546]. As illustrated in Chapter 3.2.2, the HID stands out as a unique fertile area in the northern China region because it is nourished through an artificial canal system and therefore suitable for the cultivation of vegetables and crops.

During the different periods of land reclamation beginning under the Qing Empire and reaching their peak under the governance of Mao Zedong (毛泽东) between 1949 and 1976, the expansion of cultivated land moved further northwest into the dry and unfertile areas of Inner Mongolian grasslands. Since the early nineteenth century changes in land-use rights supported land reclamation for croplands or pastures, resulting in a long-term impact on the natural environment.[547] Today, both grasslands and farmlands are affected by deterioration.

Early phase of degradation

Collecting fuel wood and other biomass to secure energy supply in rural areas were responsible for a large-scale deforestation all over China. Chapter 4.2 and 5 have illustrated and referred to the large-scale deforestation during the *Great Leap Forward*, when "backyard furnaces" in areas where coal was unavailable, were fired by fuel wood. Cutting down the trees in a concentrated countrywide deforestation campaign, gave the Leap the nickname of one of the "three great cuttings" [*san da fa* 三大伐]".[548] The second and third "cut" followed during the Cultural Revolution and later during the uncertain time of economic reforms after 1978 when land was again redistributed among the general population.[549] The consequences of these mass cuttings included erosion, sedimentation, desertification, and changes to the microclimate, which according to Shapiro, "result[ed] in a protracted loss of arable land."[550]

In addition to these large-scale wood cuttings, the over extermination of harmful insects and rodents which deeply impacted the eco-system should be also mentioned as major degrading factors during the second half of the twentieth century. Under the program of the "Four Pests" rats, sparrows, flies, and mosquitoes were marked for elimination.[551] Although their eradication was

---

546   Du, Jingyuan 2011: Irrigation Society in China's Northern Frontier, 1860's-1920's, 10.
547   Brogaard, Sara 2002: "Rural reforms and changes in land management and attitudes: A case study from Inner Mongolia, China" in: Ambio, 3, 219–225.
548   Shapiro "Mao's War Against Nature", 80.
549   Ibid.:80
550   Ibid.:80
551   Shapiro "Mao's War Against Nature", 86.

encouraged with the popular slogan, "Wipe out the Four Pests!" [*chu si hai* 除四害], the uncontrolled hunting of some species disturbed the naturally occurring balance between predator and prey.[552] Unlike the destruction of forests or the degradation of arable land, the campaign against sparrows in particular demonstrated the wider effects caused by the targeting of a specific species. Because the sparrows were killed and nearly extinct, with their removal locust populations increased rapidly and destroyed grain harvests.[553] The "Four Pests" campaign induced a major ecological imbalance rather than its intended desire to eliminate unwanted insects and other creatures from the ecosystem. In the end, it was at least partly responsible for the Great Famine of 1959 to 1961, which led to the death and suffering of millions of Chinese[554].

By 1950 China's grassland had already shrunk by 50%, whereas the number of livestock had quadrupled.[555] After 1961, the population continued to increase rapidly, but the available pasture land necessary to support it had shrunk significantly. The *Great Leap Forward* policies supporting the expansion of agricultural land in China's pastoral regions had led to a reduction of the productive rangeland for pastoralists. With the *Great Leap Forward*, Mao both encouraged and forced farmers to cultivate grassland unsuitable for raising the grain production, grain which would have been necessary to reach the maximum level to sufficiently feed the Chinese people. Mongolian leaders who complained about this campaign were arrested and downgraded.[556] The lack of power to push against these devastating practices meant that productive grassland available to minority pastoralists continuously shrunk during these campaigns. Today, largely foreign scholars are convinced that these campaigns accelerated the destruction of China's grasslands, forests, and wetlands.[557] What the Maoist Era demonstrates, is that severely transformation through man-made interventions can lead to massive and long-lasting environmental damages. In this case, of the estimated

---

552   Jiang, "The Ordos Plateau", 71.
553   Shapiro "Mao's War Against Nature", 88.
554   For further information on the Great Famine and direct influence of the Four Pests Campaign, see: Dikötter, Frank 2010 Mao's Great Famine: The History of China's most devastating catastrophe 1958–1962, 188.
      Summers-Smith, J. Denis 1992 *In search of sparrows*, 120–124.
555   Williams, "Beyond Great Walls", 29.
556   Ibid.:29
557   Smil, Vaclav. "Land Degradation in China: An Ancient Problem." In *Land Degradation and Society*. Edited by P. Blaikie and H. Brookfield, 214–22. Taylor & Francis, 2015.

sixty-seven million hectares of fertile rangeland converted to grain cultivation, only eight million hectares "survived" as grassland. The rest of these lands became cultivated fields and were exposed to degradation.[558] While the human intervention in nature led short-term to higher productivity, it also resulted in an increase in ecological damage. Despite knowing that deteriorated land cannot be used for agriculture, these short-term interventions led to the exploitation of the last prosperous areas in China for farming and depleted any possible land reserves for future generations.

Environmental degradation today

Today Inner Mongolia's environmental degradation is not limited to one particular area or system but is actually quite diverse. Across the region there is deforestation along mountains, grassland deterioration on the Mongolian Plateau, siltation and gully erosion alongside hill-slopes and desertification caused by droughts and irrigation projects as well as farmland deterioration due to agricultural mismanagement. Furthermore, like the rest of China, Inner Mongolia is confronted with environmental problems like air and water pollution due to industrialization and an ever-growing population. Inner Mongolian land-use varies across grassland types and along the transition zone between the fertile agricultural south and the dry north of the grazing lands, which are continuously drifting further north due to the increasing migration of farmers to the region[559]. The ratio of pasture land to livestock in Inner Mongolia has been in constant flux since the 1950s, and Inner Mongolia has been battling with the damage of overgrazing by livestock since the era of collectivism.[560] The typical cycle for raising livestock in the region accepts that some seasons will have droughts, while others will be more prosperous. Thus this cycle works with the natural increases and decreases in the animal population[561]. But with a sudden increase in livestock, the natural pastures are not allowed sufficient time to re-grow and the remaining grasslands have become overcrowded by the large numbers of livestock.[562] Overgrazing typically occurs when livestock numbers increase disproportionally to the amount of land available for grazing. Still it remains questionable as to whether pastoralism is as sustainable and environmentally friendly as it is so

---

558 Williams, "Beyond Great Walls", 30.
559 Ibid.:18
560 Ibid.:18
561 Ibid.:68
562 Ibid.:68

often propagated to be.[563] The deterioration of fertile grassland has definitely increased with herds that are kept too long on the same pastures. Sheep and goats ruin the top soil with their hooves and Kashmir goats eat the grass completely down to the roots, which is even worse, because it does not allow the vegetation to recover[564]. The resulting desertification and erosion leaves wasteland behind. With such major effects, can pastoralism truly be considered sustainable or environmentally friendly?

Overgrazing can cause degradation but the effects of grazing should be studied separately from the impact of cultivation on its own. It should also be taken into consideration whether livestock is the main driver of an economy or a system in addition to agriculture, a combination economy known as agro-pastoralism. The increasing population of China demands an increase in the production of red meat, which in turn has led to more intensive animal husbandry. This means an increase in livestock grazing and hay production as colder winter months require more crop cultivation in order to feed the expanding herds. Because of this interconnected relationship, it is very common to maintain cultivation and pastoralism as complimentary income opportunities in Inner Mongolia[565]. This combination of intensive farming-grazing in Inner Mongolian puts additional pressure on the existing fertile pastures and natural grasslands, which will lead to their further decline. Thus the obvious conclusion is that most of Inner Mongolia's environmental degradation should be attributed to the overexploitation of natural resources for nearly four decades, rather than climate changes over a longer period of time. Even though Inner Mongolia is dominated by unfavorable geological and geographical conditions, which are generally unsuitable for intensive agriculture its fragile ecology is almost predisposed for degradation.

Today China's policy efforts are often targeted at stopping the ongoing deterioration. The most common methods utilized to do this are by declaring new

---

563   Often traditional nomadism/pastoralism is discussed under the aspect that nomads respect their environment and live in a sustainable relation with the grasslands. But to assume that nomadic people didn't have any impact on their environment is false. Every human society, are they hunter-gatherers, nomads or peasants have transformed their environment to a certain extent with deep impact for the ecosystem.
564   For further information see article *"Mongolia: Herders caught between Cashmere and climate change"*. http://www.eurasianet.org/node/65506.
565   Ellis, J. E. et al. "Dimensions of desertification in the drylands of northern China." In *Global desertification: Do humans cause deserts? [Report of the 88th Dahlem Workshop on Global Desertification: Do Humans Cause Deserts? Berlin, June 10–15, 2001]*. Edited by J. F. Reynolds, 167–90. dwr 88. Berlin: Dahlem Univ. Press, 2002, 176.

nature reservoirs and natural parks, and reducing grazing land for agricultural use as well as for large-scale herding. Indeed, the country's ecological degradation has become one of the major factors limiting the socio-economic development of Inner Mongolia. Areas affected by severe desertification cover approximately one third of the country, and 38.2 percent of lands are seriously eroded or affected by deforestation, resulting in the degradation of natural forests and a loss of biodiversity[566]. Currently, 90 percent of China's usable grasslands are considered "degraded" which is reflected in changes in plant species and productivity loss (*Ministry of Agriculture 2007*)[567]. Specifically, over-population and over-cultivation have also been pin-pointed as the main reasons for soil erosion and deforestation. The transformation of slope-land and hilly land into cultivation areas has largely been responsible for the previous mentioned serious erosion, loss of soil fertility, and biodiversity in Inner Mongolia. To target these problems in the latter half of the twentieth century and until now, the government has launched several programs to prohibit the cultivation of crops on slope land and ecologically fragile areas. Five soil conservation methods are promoted by the Chinese government: planting woodland, grasses, alfalfa flowers, building contour earth banks, and terraces.[568] Other factors have also prompted government reaction, such as the serious Yangtze, Songhua and Neijing River basins flood in 1998, which led to the first introduction of the *Grain for Green* (GfG) program in three provinces, Sichuan, Shaanxi and Gansu[569]. This program sought to increase income levels in the rural population in such a way that the subsidies given for converting farmland into forestland were higher than the typical net-income for grain production on sloping land[570]. Thus it made environmentally-positive action more profitable for the farmers in these slope regions, and led to a decrease of slope-land farming. Attempts were also and are still being made for reforestation and driving back the expansion of deserts and eroded tracts of land. According to Delang and

---

566 Liu, Can. "Grain for Green Program in China: Policy making and implementation?", 4.
567 Wenjun Li, Lynn Huntsinger 2011: "China's Grassland Contract Policy and its Impacts on Herder Ability to Benefit in Inner Mongolia: Tragic Feedbacks". In: Ecology and Society, 16 (2), 1–14.
568 Zhu, T. X., and A. X. Zhu. "Assessment of soil erosion and conservation on agricultural sloping lands using plot data in the semi-arid hilly loess region of China." *Journal of Hydrology: Regional Studies* 2 (2014): 69–83. doi:10.1016/j.ejrh.2014.08.006.
569 Liu, Can. "Grain for Green Program in China: Policy making and implementation?", 2.
570 Ibid.:11

Zhen (2015), it appears that the GfG policy did have some early success in slowing country-wide environmental destruction as well as in combating deforestation, ecological degradation, over-grazing, soil erosion, and the over-cultivation of sloping land.[571] Nevertheless there is no nation-wide assessment on the latest impacts of the GfG program and its results are still questionable, whether the overall outcome of this program has positively influenced China's further environmental development.[572] It would seem that when the government decides to enact specifically targeted environmental improvement policies, success can be achieved. In the following section, two major environmental problems in different areas of Inner Mongolia will be presented together with an explanation of the methods and measures that are being implemented in an attempt to reduce the chronic water-shortage facing the region and the progressive destruction of its eco-environment as well as the expansion of deserts in Inner Mongolia.

7.1 Irrigation and Water Shortage in Inner Mongolia – A Severe Problem for Farmers and Herdsmen

The total arable land in north China accounts for 65 percent of land in China, but the water resources are less than 20 percent of whole of the country, whereas the fertile South covers only 35 percent of arable land, but commands of 80 percent of the water reserves[573]. This natural distribution of little rain-fed fertile areas, huge drylands and large deserts (see Map 7) are together a formidable challenge for water-saving agricultural practices. It is assumed that Chinas drylands will further expand and make sustainable agriculture more difficult. Therefore the author is fully convinced that only by improving the water-saving irrigation and adopting new, sustainable water-saving techniques in the matter of transportation and application methods are the only solution to sustain dryland agriculture in Inner Mongolia.

---

571 Ibid.:2
572 Delang, Claudio O., and Zhen Yuan. *China's grain for green program: A review of the largest ecological restoration and rural development program in the world*. Cham: Springer, 2014, 135.
     http://search.ebscohost.com/login.aspx?direct=true&scope=site&db=nlebk&db=nlabk&AN=888739.
573 Ibid.:1

Map 7: Precipitation Distribution China

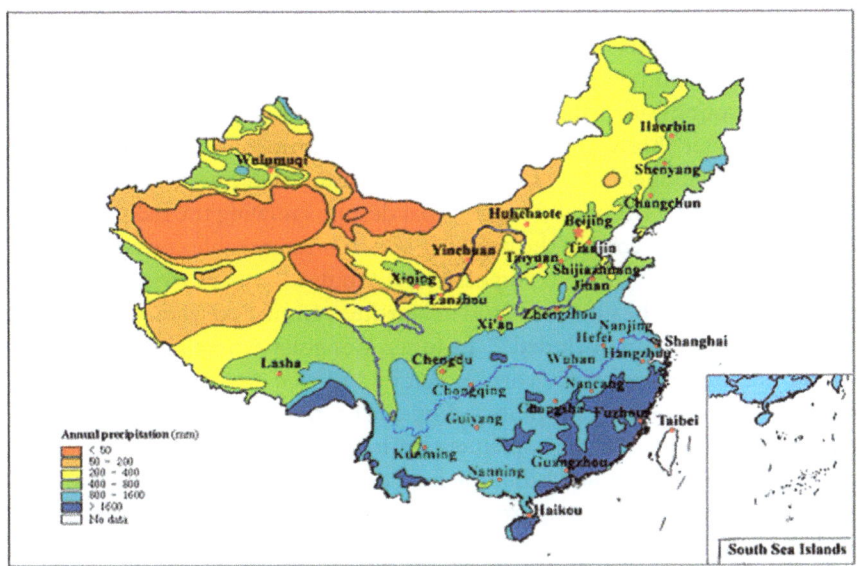

Source: China Meteorological Administration. http://www.regional.org.au/au/asa/2004/symposia/1/5/2085_deng.htm

It is the chronic water shortage, which is the greatest problem facing Inner Mongolia's dryland agriculture. This issue is also threatening China's food security, limiting the economic development and the quality of life for the area's residents. Furthermore, the battle for water resources and water-use between industry and agriculture, and between urban and rural areas is obviously a conflict of interests. In Inner Mongolia specifically, the severe water scarcity refers to both insufficient water resources and reduced water quality. Over-exploitation of water in the region has led to serious problems for agriculture: ground subsidence, salinity intrusion and ecosystem deterioration. Because a sufficient water supply depends on resources and sustainable water-management technologies the fact that the natural water resources of Inner Mongolia are characterized by scarcity and abundance of rainfall is clearly problematic. Currently, the most important water resource is the Yellow River, which nourishes nine provinces and autonomous regions. However, unsustainable and excessive water-use in Inner Mongolia and Ningxia has had a much greater impact for the water-users downstream the Yellow River in Henan and Shandong. The dangers alongside the lower reaches of the Yellow river are well-known facts. Specifically, riverbeds are drying up, and in other places sedimentation and the deposition of silt is causing the river bed to rise and overrun its banks[574].

The middle reaches of the Yellow River are partly irrigating the Hetao Plain, Ordos Plateau and Loess Plateau of Inner Mongolia. However, in the outlying areas the groundwater table is sinking, rain-fed irrigation is insufficient and the severe water shortage continues to increase. Although water shortages are a common phenomenon for this area, the irrigation water use efficiency is still only 40 percent[575]. This would suggest that despite knowing the risks, few steps have been taking to combat the problem, which is remarkable considering that maintaining food security is impossible without additional irrigation. More than 70 percent of the total farmland in Northern China is dryland agriculture, crops and pastures are included[576]. In Inner Mongolia, where major crops like wheat, oat, millet, sorghum, corn, potatoes and rice are planted and a wide range of cash crops are grown, including soy beans, linseed, rapeseed, castor-oil plants and sugar-beets, supplementary irrigation is absolutely essential in order to increase yields and fulfill the food demands of the whole nation. Taking into consideration that approximately two-thirds of China's people live in arid and semi-arid regions, it is vital that water-saving agriculture techniques be implemented in order to address the food scarcity issues and accelerate crop yields.

China's water-shortage: Salinization and infertile soils in Inner Mongolia

Today water stress is the most important issue for China to solve and control. The assumption is that over time three factors contributed to today's water stress in China:

---

574 Liu, Changming 2002: "*Drying up of the Yellow River: Its impacts and countermeasures: Mitigation* and Adaptation" in: Mitigation and Adaptation Strategies for Global Change, 7, 203–214.
575 Xi-Ping Deng et. al. "Improving Agricultural Water Use Efficiency in Arid and Semiarid Areas of China: "New directions for a diverse planet"." Proceedings of the 4th International Crop Science Congress, 26 Sep-1 Oct 2004, Brisbane. Accessed November 11, 2014. Available at: www.cropscience.org.au.
576 Ibid.:1

1. Decline in water supplies
2. Increase in demand
3. Growing awareness of environmental water needs[577]

Applying these same factors to the situation in Inner Mongolia, these three issues become even more prominent. The decline of the water supply can be attributed to both the shortage of naturally occurring water resources in the north and changes in land use which have increased water pollution and the need for artificial irrigation. The main water pollution source in this region has been unmeasured wastewater discharge from cities, industries as well as urban townships and villages. Furthermore, the increased use of fertilizers and pesticides that enter rivers and groundwater have also been partially responsible for the poor water quality. All of these factors combined have led to a decline in Inner Mongolia's water use efficiency.

Salinization is an additional threat to Inner Mongolia's agriculture, though it is important to distinguish between two distinct forms of salinization potentially affecting the area. Natural soil salinization occurs with various natural processes and develops over a longer period of time. Man-made salinization can be the result of irrigation systems and additional water-input.[578] Irrigation water from the Yellow River contains naturally-occurring salt, which is left behind in the surface soil, after the water has evaporated. Particularly in combination with a dryland environment, where natural *"salts are deposited and stored in the unsaturated zone and are eventually transported to shallow groundwater that discharges into adjacent rivers"* the salinization balance of surface waters as well as the drainage water is endangered for salinization. As A. Vengosh (2014) has pointed out: "*Once the natural salt balance is disturbed and salts begin to accumulate, either in the unsaturated zone or in drainage waters, the salinity in the downstream discharge water will increase*" too.[579]

---

577 Jiang, Yong. "China's water scarcity." *Journal of Environmental Management* 90, no. 11 (2009): 3185–96. Accessed March 13, 2015. doi:10.1016/j.jenvman.2009.04.016. 19539423.

578 Yu, Ruihong, Tingxi Liu, Youpeng Xu, Chao Zhu, Qing Zhang, Zhongyi Qu, Xiaomin Liu, and Changyou Li. "Analysis of salinization dynamics by remote sensing in Hetao Irrigation District of North China." *Agricultural Water Management* 97, no. 12 (2010): 1952–60. Accessed June 11, 2013.

579 Vengosh, A. "Salinization and Saline Environments." In *Treatise on Geochemistry: Treatise on Geochemistry*. Edited by Karl K. Turekian. 2nd edition, 325–78, 330. Amsterdam: Elsevier Science, 2013.

As mentioned in chapter 3.2.2 about the HID, which serves as an example for man-made salinization, the canal system and its flooding system has largely been responsible for an increase of salt deposits in the surface soil. Improper drainage systems are largely contributing to Inner Mongolia's soil salinization and they are representative for certain agricultural processes severing the global environmental problem of soil salinization, especially in drylands. China alone currently faces salinization and alkalization of 7 percent of its total land area, which resembles an area of 3.69×107 hm2 (hectometers)[580]. This land area is generally concentrated in the regions further inland, in the arid and semi-arid regions of north China, where natural precipitation is low, and deserts and grasslands dominate the landscape[581]. Although most of China's arable land is tendentiously slightly salinized, the agricultural production area of Inner Mongolia, especially HID is highly salinized, which has majorly impacted regional production[582]. The vast majority of irrigation water in Hetao comes from the Yellow River floating through HID canal-system, which has been discussed in 3.2.2. The main irrigation canal is 180 km long and the drainage system is 260 km. Groundwater is mainly used for domestic and industrial purpose, as well as for the pasture sector. 96% of the water usage is for irrigation and the increasing water demand for industrial, agricultural purposes, including hydropower stations aggravates the problem of water scarcity in the Yellow River basin.[583]

But cultivation of cash crops doesn't only concentrate on the above mentioned irrigated areas. Major cash crops like wheat, maize, oil seeds and sugar beets are cultivated all over Inner Mongolia, because of their high profitability for the famers. Additionally millet, sesame, melon and other fruits as well as pasture grass are planted. Because of less rainfall the entire crop season requires artificial irrigation systems to maintain the crop production[584]. But this increase in water-use

---

580 China Council for International Cooperation on Environment and Development. "Developing Policies for Soil Environmental Protection in China." CCICED Annual General Meeting Unpublished manuscript, last modified December 12, 2013. http://www.cciced.net/encciced/policyresearch/report/201205/P020120529358298439639.pdf.
581 Yu, Ruihong et al. "Analysis of Salinization", 1953.
582 CCICED Annual Report 2010.
583 Ibid.:301
584 Xu, Xu, Guanhua Huang, Zhongyi Qu, and Luis S. Pereira. "Assessing the groundwater dynamics and impacts of water saving in the Hetao Irrigation District, Yellow River basin." *Agricultural Water Management* 98, no. 2 (2010): 301–13. Accessed June 12, 2013.

for agricultural production has put pressure on the fragile ecosystem of Inner Mongolia's lifeline, the Yellow River.

Since 1990, the lower reaches of the Yellow River have had annual struggles with severe drought. However, since the 1990s, the *Yellow River Water Conservancy Commission* (YRWCC) has been working on forecasting and water management issues. To date, they maintain the stance that this problem can only be solved by a, massive reduction in agricultural irrigation[585]. In particular, the YRWCC names "excess water diversions and poor irrigation and drainage management-systems"[586] as the major culprits behind this serious issue. Thus, sustainable water management and usage at both the farm and district levels are necessary in order to reduce the risk of an imbalance in water resources available between the upper and lower reaches of the Yellow River[587]. Still, it is significant that the huge demand for river water is for agricultural purposes as an additional water resource, it is less used for industrial purposes due to its high sediment load and it is costly to filter the sand out. Furthermore, the uncontrolled usage of groundwater leads to the problem of salinization. Namely, if the groundwater table is raised due to poor drainage systems, the surplus water evaporates and leaves salt behind. Today, scholars suggest a more conscious approach to groundwater usage and Yellow River water usage in combination with effective water management policies in order to establish more sustainable development patterns and to find a way of water-usage balance[588].

Beyond droughts, soil salinity and the acidification of infertile, exhausted soils are a further problem facing Inner Mongolia. The use of organic fertilizers like compost and feces can help to improve the soil quality. To maintain water-saving agricultural practices it is essential that users take full advantage of the natural water resources and irrigation facilities. Therefore a major challenge in Inner Mongolia is how to increase the water-use efficiency for high crop yields by using a minimum of supplementary water[589]. One method to increase water-use efficiency is by shifting to a more efficient irrigation system[590]. Specifically, drip irrigation systems are more efficient than sprinkler or pivot irrigation. Border or furrow irrigation, a further system, has also been discredited for significant water waste. By actively supporting different soil management practices additional

---

585   Ibid.:302
586   Ibid.:302
587   Ibid.:302
588   Yang, "Remote sensing", 112.
589   Ibid.:4
590   Deng, "Improving Agricultural Water Use", 8.

savings can also be achieved. The use of plastic mulch, which is a sheet of plastic used to protect soil evaporation, is a common method, and is often used in potato cultivation[591]. Slope terracing and contour farming is widespread in the semi-arid Loess Plateau of Inner Mongolia, where the farming is also affected by gully erosion. These methods have shown that terracing in the Loess Plateau improved water infiltration, increased the "rainfall utilization rate" and achieved higher crop yields[592]. Terracing is widely spread in areas with gentle gradients. Terraced fields prevent surface water run-off and soil erosion. Nevertheless the author sympathizes with the view of Brian Finlayson (2002) that most of the terracing in the Loess Plateau and slope areas of Inner Mongolia are constructed to cultivate unproductive farmland rather than reducing sediment flow into the Yellow River.[593]

Development of China's water management

The history of water development and management has been always been both an important aspect of Chinese life and a national priority to maintain the diverse canal systems utilized for grain transport to the northern capitals[594].

During the nineteenth century, the control and regulation of water resources continued. By the twentieth century, these practices turned toward modern

---

591 For further information on the application of plastic mulch on potato fields and its impact on water-use efficiency please see:
Hou, Xiao-Yan, Feng-Xin Wang, Jiang-Jiang Han, Shao-Zhong Kang, and Shao-Yuan Feng. "Duration of plastic mulch for potato growth under drip irrigation in an arid region of Northwest China." *Agricultural and Forest Meteorology* 150, no. 1 (2010): 115–21. Accessed November 20, 2014.
Wang, Feng-Xin, Xiu-Xia Wu, Clinton C. Shock, Li-Yun Chu, Xiao-Xiao Gu, and Xuan Xue. "Effects of drip irrigation regimes on potato tuber yield and quality under plastic mulch in arid Northwestern China." *Field Crops Research* 122, no. 1 (2011): 78–84. Accessed November 20, 2014.
Wang, Feng-Xin, Shao-Yuan Feng, Xiao-Yan Hou, Shao-Zhong Kang, and Jiang-Jiang Han. "Potato growth with and without plastic mulch in two typical regions of Northern China." *Field Crops Research* 110, no. 2 (2009): 123–29. Accessed November 20, 2014.
592 Deng, "Improving Agricultural Water Use", 9.
593 Finlayson, Brian. "Managing Soil Erosion on the Loess Plateau of China to Control Sediment Transport in the Yellow River-A Geomorphic Perspective." Unpublished manuscript. http://www.tucson.ars.ag.gov/isco/isco12/VolumeI/ManagingSoilErosionontheLoessPlateau.pdf.
594 Pietz, D., and M. Giordano. "Managing the Yellow River: continuity and change." In *River basin trajectories: societies, environments and development*. Edited by F. Molle and P. Wester, 99–122. Wallingford: CABI, 2009:104.

hydraulic engineering with mass mobilization during Mao Zedong's leadership, resulting in the construction of huge river dams and hydropower plants.[595] After the death of Mao in 1976, a new era in water management began. With the abolishment of the commune system, replaced by the HRS in 1984, the government control became increasingly decentralized. During this period the first signs of environmental awareness began to appear, cumulating in the 1988 Water Law, which provided the basic framework for water management in the 1990s.[596] This law was further revised and renewed in August of 2002[597].

Throughout this time, concepts for water pricing, regulating water rights and water markets have been discussed for their positive influence on the water-use behaviour of the residents. In particular, the debate about pricing water through markets is an ongoing discussion between development and aid agencies. The Asian Development Bank as well as the World Bank have both advocated the allocation of water through markets by using the system of water rights supported by legal and institutional frameworks as a solution to long-standing problems connected to water-use efficiency and access to water.[598] Furthermore, the YRWCC's plan since 1999 has been to manage the water scarcity crisis through engineering, raising water prices, establishing a system of water rights transfer and by supporting research based on similar concepts from Australia and the United States.[599] In general, the water prices charged to farmers in China consist of a resource fee and infrastructure charge set by the state. Improving rural incomes is an important part of China's current national policy for increasing economic growth. Still, the government has been struggling with the issue that increasing the price of water can be a sufficient tool to save water and make further investments possible. At the same time, the raising of prices is threatening farmers' welfare, who are the largest water-user group. The Water Law (2002) clearly defines water resources as belonging to the state, with the State Council holding ownership over these rights. The water resources in ponds and reservoirs, which were built by the rural collective organizations, can be used without permission

---

595 Ibid.:108
596 Ibid.:110
597 Water law of the People's Republic of China. Accessed January 2016. Available at: http://www.china.org.cn/english/government/207454.htm.
598 Webber, Michael, Jon Barnett, Brian Finlayson, and Mark Wang. "Pricing China's irrigation water." *Global Environmental Change* 18, no. 4 (2008): 617–25.
599 Ibid.:618

for agricultural or daily purposes.[600] Thus the current discussion on changing the water rights in China would mean significant changes in laws connected to usage rights as the rights to water are owned by the Chinese government.

7.1.1 The Inner Mongolian Water-Rights System and the Function of Water User Associations

The system of water-rights in Inner Mongolia is basically the right of water resource utilization. The Water Law of the PRC (2002) clearly defines that the water resources, either rivers or lakes, belong to the State and the State Council on behalf of the PRC, which holds the ownership.[601] In Inner Mongolia, water rights are more precisely a water-use permit. Industrial users are required to obtain this water-use permit, which is valid for five years, from the local government after passing a water-resource assessment. In exchange, the fee for the water-use permit is paid to the local government, which re-invests the funds in agricultural water-saving projects[602]. The rural collective can use the water without permit but only for daily or agricultural purpose. Currently, the main problem in Inner Mongolia is that groundwater is not included in the process of water-right transfers, and thus the groundwater extraction is not clearly regulated. This increases additional private drills, which are utilized to pump the groundwater and are responsible for sinking groundwater tables. As a consequence, the government heavily promotes drip irrigation and partially allows drilling, but not in regions where the water table has already decreased, as the Head of the Agricultural Bureau in Hohhot admits.[603]

Even though the Chinese government has encouraged implementing water-saving technologies since the 1990s, the adoption rate in rural villages has been slow, especially for technologies like plastic mulch, sprinkler irrigation systems and drought resistant varieties of crops, with the exception of the potato[604]. Water-pricing as discussed previously, has also made little progress. In general, the small-scale and fragmented farming conditions of Inner Mongolia

---

600 Jia, Shaofeng 2014: Water rights in China. Accessed January 2016: available at: http://chinawaterrisk.org/interviews/water-rights-in-china/.
601 Ibid.
602 Ibid.
603 Confidential information from the Interview with the Head of the Agricultural Bureau in Hohhot, see appendix vi.
604 Huang, Qiuqiong, Scott Rozelle, Jinxia Wang, and Jikun Huang. "Water management institutional reform: A representative look at northern China." *Agricultural Water Management* 96, no. 2 (2009): 215–25. Accessed November 4, 2014.

complicate the measuring of water and volumetric pricing[605]. In combination with the goal to further poverty alleviation through the elimination of taxes for the rural population, resistance against the theory of water-pricing is growing. This is largely due to the fact that for farmers the actual pricing by land-unit is much more profitable than the proposed pricing by volumetric charges.

One other efficient method to alleviate water supply problems and poverty are Water User Associations (WUA). WUA also have their origins in the 1990s, when this model was being heavily promoted by the World Bank, as based on positive experiences in other countries. As the theory goes, the participation of peasants in the water management and regulation process helps to reduce the financial burden on the government, optimizes water-use efficiency and supports the operational efficiency of hydraulic water facilities, thus minimizing water shortages[606]. In Inner Mongolia, WUA are being promoted in order to secure the farmers' harvests, regardless of precipitation. As a consequence, the farmers' incomes will typically increase, leading to a slight improvement in individual poverty levels. In 2006, 2,876 water organizations existed in Inner Mongolia, of which 519 were WUA[607]. Since 2009, many more WUA have been established in Inner Mongolia in order to assist in the regulation of large irrigation districts.[608]

The main purpose of a WUA is to organize a group of farmers into their own non-profit cooperative in order to maintain and manage the water delivery facilities within their region[609]. WUA are especially useful in areas with large irrigation districts (ID). For example, the HID, where farmers select leaders and establish a set of rules for the management of water deliveries purchased directly from the ID on a per unit basis. At the same time, individuals are selected to maintain the selected canals and receive governmental incentives for distributing the water efficiently[610]. With this reform, the bureaucracy associated with fees is transferred from the village-township-county level to the ID, thus reducing the

---

605 Ibid.:216
606 Qiao, Guanghua, Lijuan Zhao, and K. K. Klein. "Water user associations in Inner Mongolia: Factors that influence farmers to join." *Agricultural Water Management* 96, no. 5 (2009): 822–30.
607 Ibid.:823
608 Ibid.:823
609 Bryan Lohmar, Jinxia Wang, Scott Rozelle, Jikun Huang, and David Dawe. "China's Agricultural Water Policy Reforms: Increasing Investment, Resolving Conflicts, and Revising Incentives." *Agricultural Information Bulletin*, no. 28 (2003): 1–34. Accessed April 1, 2015: 1–34.
610 Ibid.:20

fees overall and saving money for farmers. The collection of fees then supports the local ID's fiscal situation relating to water use.[611] The knowledge exchange between different WUAs is also advantageous, as new water-saving technologies or practices can be introduced and implemented. It has also been shown that these systems are much more easily adopted by farmers because they are forced to be more aware of their water use and costs, as the fees for water use are collected separately and not in combination with other taxes or fees under the previous system.[612] Still, the World Bank recommends five principles that should be followed in order to guarantee agricultural water use efficiency and ensure the effectiveness of a WUA (Tab. 10).

Tab. 15: Principles of Water User Associations

5 PRINCIPLES OF EFFECTIVE WUA (described by Worldbank)[613]	
1	Adequate and reliable water supply & good delivery infrastructure
2	Farmer's organization & elected leadership
3	Jurisdiction within the hydraulic boundaries
4	Contracts with water suppliers & measuring in volumetric prices
5	Assessing & collecting water charges from its members

Source: Cilia Neumann 2016, based on Julian Doczi, Roger Calow and Vanessa d'Alancon (2014): *Growing with less. China's progress in agricultural water management and reallocation. Case study Report Environment.* In: *Development Progress.*

It is clear that sustainable water-use management is very important to reduce salinization and alkalization in Inner Mongolia's agricultural farmland where the already diminished soil fertility has led to a loss in crop production. The growing environmental awareness has helped to establish private initiatives like WUA, where local people try to work sustainably and actively take part in reducing harmful practices. In combination with other sustainable methods, such as crop-rotation or herd shifting as well as planting trees and other plants to prevent soil erosion, modest accomplishments can definitely be achieved.

---

611 Ibid.:20
612 Ibid.:21
613 Julian Doczi, Roger Calow and Vanessa d'Alancon. "Growing With Less: China's progress in agricultural water management and reallocation." *Development Progress*, 2014; Case study Report Environment. Accessed April 13, 2015.

7.1.2 Case Study 2: Irrigation in Siziwang Qi Banner

The following case study is based on the second part of the Siziwang questionnaire, which was focused on environmental issues confronting farmers in Inner Mongolia. Therefore the questionnaire was centered on the question: "Do the farmers perceive their environment as intact or do they fear any environmental degradation?"

Being hit by drought or water shortage was a main issue for the farmers in Siziwang Qi. As previously mentioned, Siziwang Banner land-use is mainly dedicated to cultivated farmland and the distribution of crops is based on staple foods and oil-producing plants.

Tab. 16: Crop Distribution among Thirty Farmers in Siziwang Qi

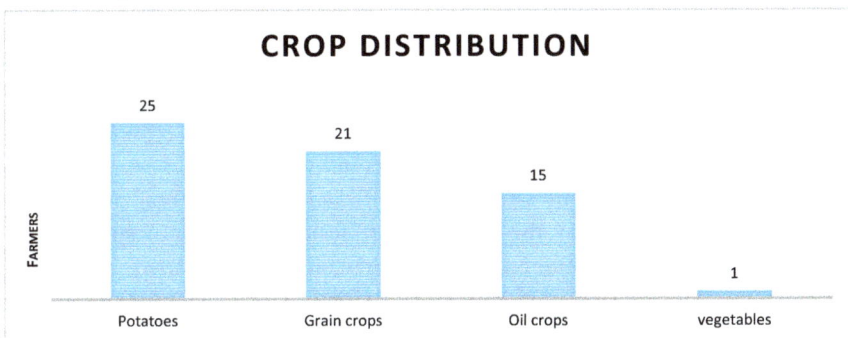

Source: Cilia Neumann 2016, (based on field research Inner Mongolia, September 2013) *"Distribution of cash crops in Siziwang Qi"*

The majority of the farmers interviewed confirmed that despite the associated issues with proper irrigation, they still invest in and support the cultivation of staple crops like potatoes and grains (millet, buckwheat, corn and wheat). In contrast, the cultivation of oil crops (rapeseed, sunflower) has increased largely due to the high efficiency of these plants in terms of water usage. The fact that oil crops increase the income of the farmer's household because the profitability rate of these cash crops is much higher tends to strengthen the decision-making process in the direction of farmers favoring the cultivation of potatoes over wheat. Nevertheless, the cultivation of wheat is deeply rooted in the Chinese tradition of grain self-sufficiency and which might additionally help to explain why it is still wide spread, despite the state "control[ing] grain production, processing, storage, as well as internal and external trade."[614]

Detailed evaluation shows that 83.3 percent of the farmers plant potatoes, 70 percent plant grain crops, 50 percent plant oil crops and 3.3 percent cultivate vegetables. None of the farmers questioned support monoculture in their production, but rather they consistently support crops in combination with each other. Most predominantly seen in Siziwang Qi is the combination of potatoes and a grain or grains (70 percent), potatoes and oil seed (60 percent), or grains and oil seed (56.6 percent), or a combination of all three. Still, in terms of water use efficiency, low use of inputs and poor land quality, grain and oil seed show little economic viability under these conditions when compared to potatoes. Grain and oil crops are much more reliant upon water and need sufficient irrigation in order to bring in profitable revenues in terms of yield and income. With regard to water-shortage and irrigation issues, 66 percent of the people interviewed confirmed that they are affected by drought and lack of sufficient water. Therefore planting potatoes makes sense both in regard to higher yields while simultaneously depending on reduced water-input. It is both the most economical and environmentally-sound choice for the farmers in this region.

Although, the farmers in Siziwang are affected by the general problem of soil erosion and salinization, the decline in soil fertility is a major issue in Inner Mongolia. Even in the fertile region of the irrigated Hetao district, where the Yellow River and the irrigation canals are partially responsible for the loss of soil nutrients, as well as in the northern dryland plains and plateaus, which are characterized by gully erosion and desertification, farming is very difficult. In Siziwang Qi, 36 percent of farmers blame the poor soil quality (eroded, highly salinized and alkalinized) for the bad harvests and the difficulties they face in achieving profitable farming results.

---

614 Zhang Hongzhou 2015: *"Can the potato help feed China, cut pollution and alleviate drought?"* https://www.chinadialogue.net/article/show/single/ch/7657-Can-the-potato-help-feed-China-cut-pollution-and-alleviate-drought-

Tab. 17: *Soil Issues among Thirty Farmers in Siziwang Qi*

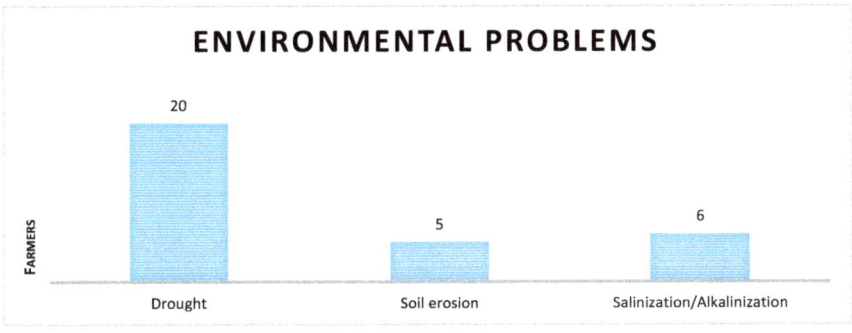

Source: Cilia Neumann 2016, (based on field research Inner Mongolia, September 2013) "*Soil issues affecting farming profitability in Siziwang Qi*"

Of the questioned farmers, who are mainly hit by drought, 45 percent blamed climate change and/or weather conditions for the insufficient irrigation and water shortage. Nevertheless, the reasons for inefficient water supply as well as use are much more diverse. Beside the excessive use of ground water and diverted rivers, which cause the lower reaches to dry up, it is mainly the wasteful use of water that is responsible for the severe problems in Siziwang Qi. Leaking water pipelines and flood irrigation that seep cubic meters of water on a daily basis are wide-spread. One solution to combat this problem would be the transition to drip-irrigation systems or other efficient water supply systems, which are substantially better in their water use and would help significantly with water shortages.

Although 33.3 percent of the farmers explained that promoting water saving irrigation technologies which are sustainable and the most reasonable method to prevent the chronic water shortage and irrigation issue, only 16.6 percent actually use their re-investment for water-saving technologies like drip irrigation or plastic mulch. Of the farmers questioned, 46.6 percent preferred to invest their money in machines with the desire to increase work efficiency through technological methods like mechanical planting, harvesting and sorting. However, 30 percent of the farmers questioned had made no attempt to re-invest in any kind of technological improvement (Tab. 18). As mentioned in chapter 6.3 seed care and pesticide treatment seem to be of little interest for the farmers in Siziwang Qi, which is contradictory to the outcomes provided by Yun Ting, owner of the *Zhengfeng Seed Potato* company as well as the Head of the Agricultural

Bureau in Hohhot, claiming that the seed tuber market is "the key for the future to increase the yield"[615].

Tab. 18: *Allocation of Re-Investment among Thirty Farmers in Siziwang Qi*

RE-INVESTMENT

Category	Farmers
Machines	14
Irrigation	3
Seeds/Pesticides	2
Greenhouse/Plastic mulch	2
No investment	9

Source: Cilia Neumann 2016, (based on field research Inner Mongolia, September 2013) *"Farmers' allocation of re-investment"*

It should also be noted here that the proper storage of crops, especially the potato, is a major issue in Inner Mongolia. Most of the yield rots due to improper storage practices, which lead to fungal growth or other diseases and pests. Therefore investments in better storage methods and the enlargement of storage capacity are both very important to maintain a sustainable and profitable crop production. It is thus essential to educate the farmers in Siziwang about sustainable investments they can make and supporting them in finding the correct strategies and solutions to improve their cultivation methods. Here international partnerships between agri-businesses and non-governmental organizations, like the WTO and the CIP in Beijing, are initiating diverse programs to support smallholders in adapting to global markets.[616] The Small Farmers Adapting to Global Markets (SFAGM) was a program from 2003 to 2008 that was initiated by the Canadian and Chinese governments and carried out in Inner Mongolia and Sichuan.[617] The program's focus was on four main aspects, which were to improve the "Food Safety System, [enhance] the WTO Compliance Capability

---

615 Confidential information from the Interview with the Head of the Agricultural Bureau in Hohhot, see appendix vii.
616 SFAGM, *Small Farmers Adapting to Global Markets*; Syngenta *Good Growth Plan*.
617 Zhang, Lei, and Meiju Wu. "China-Canada "Small Farmers Adapting to Global Markets" Project." http://www.oecd.org/aidfortrade/47027097.pdf.

(Capability Building), [implement an] Agricultural Administration Reform (Policy Research) and [adapt] Small Farmers to the Markets in Conformity with WTO Rules (Pilot Project for Provinces in Western China)."[618]

Drought and water-shortage are absolutely manageable environmental problems, as the farmers in Siziwang have demonstrated. But the question remains: How should the people deal with expanding sand dunes and deserts burying farmland and pastures? This question is the subject of the following section.

7.2 Desertification and Sandstorms – A Threat to the Local Population

According to the Convention to Combat Desertification and Drought in 1994, desertification is defined as "land degradation in arid, semi-arid, and dry sub-humid areas resulting from various factors including climatic variability and human activities."[619] As Reynolds and Smith explain, there is contentious debate surrounding the term 'desertification' because this definition implies that rather than being natural phenomena, deserts are created by something or someone. The general assumption they present is that climate has a controlling function on ecological systems and may influence desert expansion, but that humans and their practices, either agriculture or the ranching of animals, have both attributed to "causing" desertification in some places. Still scientists do not agree about to what extent these components are responsible for desertification. Although answering this question is not subject of this chapter, it is important to keep in mind that the factors connected to the desertification process are both multifaceted and disputed.[620]

Climate change is a global phenomenon and should of course be considered as partially 'responsible' for the changing ecological circumstances in Inner Mongolia. As Mengli Zhao explains, "the climate of Inner Mongolia is getting drier and windier" and precipitation is further declining.[621] In combination with

---

618 Ibid.:1
619 Chen, Y., and H. Tang. "Desertification in north China: background, anthropogenic impacts and failures in combating it." *Land Degradation & Development* 16, no. 4 (2005): 367–76. doi:10.1002/ldr.667.
620 Reynolds, J. F. and Stafford Smith, D. M, eds. *Global desertification: Do humans cause deserts? [Report of the 88th Dahlem Workshop on Global Desertification: Do Humans Cause Deserts?, Berlin, June 10–15, 2001]*, 88. Berlin: Dahlem University Press, 2002.
621 Zhao, Mengli 2006: "Grassland Resource and its Situation in Inner Mongolia, China" in: *Bull.Facul.Agric.Niigata Univ.*, 58 (2) [新潟大学農学部研究報告, 第58巻 2 号], 129–132.

global warming and increasing temperatures, the starting situation for fertile soils is already disadvantageous. Extreme droughts in Inner Mongolia, like those between 1999 and 2001, have further destroyed the native grassland and farmland. The destruction or decline of these eco-environments has furthermore led to a decrease in wildlife species of plants and animals. Due to the extinction of predators the numbers of pests and rodents have also increased and are a threat to crop harvests. Locust infestations are a common phenomenon of degraded lands where the soils have high nitrogen concentrations.[622] To summarize, Inner Mongolia's physiographic conditions, which are characterized by an inherently harsh eco-system with barren vegetation, a continental climate and sandy soils are largely predestined for desertification. Inner Mongolia's climate, specifically as part of north China, has demonstrated climatic fluctuations between 1960 and 1990 with a growing tendency towards aridity.[623] Other studies reported a general rise in temperature in the region.[624] For Inner Mongolia the rainfall records indicated that the annual precipitation between 1963 and 2000 was constantly under the previous norms.[625] The deserts and sandy lands in Inner Mongolia are dominated by moving sand dunes, which are in constant migration towards the south-east due to wind drifts from the north-west. This lack of rainfall has led to a paucity of sufficient ground and surface water in Inner Mongolia.[626] The shortage of river runoffs in north China has been problematic because despite being one of the primary water sources some, like the Haihe River, had shown "zero flow for 20 out of 26 years from 1972 [to] 1997", meaning that there was no water runoff for several years[627] The drying up of rivers leads to land aridity and a dropping of the groundwater levels, which also contribute to the development of deteriorated and desertificated landscapes. In summary, the impact of climatic factors including temperature, precipitation and sand storms may have a much stronger effect on desertification than previously assumed.

---

622 Cease, Arianne J. 2012: "Heavy livestock grazing promotes locust outbreaks by lowering plant nitrogen content" in: Science (New York, N.Y.), 335, Nr, 6067, 467–469.
623 Chen and Tang, "Desertification in north China", 369.
624 Chen L., Zhu W., Wang W., Zhou X., Li W. 1998: "Studies on climate change in China in recent 45 years". In: Acta Meteoroligica Sinica 56, p. 257–271.
Shang K., Dong G., Wang S., Yang D. 2001: "*Response of climatic change in north China deserted region to the warming of the Earth*". In: Journal of Desert Research 21 (4), p. 387–392.
625 Chen and Tang, "Desertification in north China", 369.
626 Ibid.:370
627 Ibid.:370

Even though some scientists believe that the impact of human activity on environmental changes in north China has been overestimated, the influence of overgrazing and agriculture in Inner Mongolia is obvious and has been investigated by several scholars[628] The investigation of drylands and deserts in China began in the late 1950s in parallel to the control and land reclamation in the North by the Chinese government. The long process of urbanization and industrialization (1960–1990) which accompanied this control had a major effect on China's northern steppe ecosystem. Since then a major decline in the loss of fertile soils all over China has been recorded[629]. Various processes of land degradation have led to large areas of unusable land. The remaining fertile lands were further isolated by the increasing desertification of surrounding areas. On these "fertile islands" the production of agriculture or livestock intensified, and therefore led these areas also becoming degraded[630]. During the last one hundred years, Inner Mongolia has undergone two periods of increase in livestock production when the animal density exceeded the rangeland capacity between 30 and 70 percent, from 1950 to 1965 and from 1985 to 2000. These periods of intensive overgrazing diminished the vegetation, exposing it to wind erosion and desertification. By 2005, the population of Inner Mongolia had grown fourfold since 1950, and the livestock raised in the region had increased from 10 million to 60 million within the same time period, resulting in the increases of overgrazed pastures, which in turn have

---

[628] Wang, Xunming; Chen, Fahu; Dong, Zhibao 2006: "The relative role of climatic and human factors in desertification in semiarid China". In: Global Environmental Change 16 (1), p. 48–57.
Wang, Xunming; Chen, Fahu; Hasi, Eerdun; Li, Jinchang 2008: "*Desertification in China. An assessment*". In: *Earth-Science Reviews* 88 (3–4), p. 188–206.
Yang, X.; Ding, Z.; Fan, X.; Zhou, Z.; Ma, N. 2007: "Processes and mechanisms of desertification in northern China during the last 30 years, with a special reference to the Hunshandake Sandy Land, eastern Inner Mongolia". In: CATENA 71 (1), p. 2–12
Meizhen Liu, Gaoming Jiang et al. 2002: "*The control of land degradation in Inner Mongolia: a case study in Hunshandak Sandland*". (Ed.) UNESCO-MAB. Chinese Academy of Sciences. Beijing (KSCX1-08-02).
Zhao, H.-L.; Zhao, X.-Y.; Zhou, R.-L.; Zhang, T.-H.; Drake, S. 2005: "*Desertification processes due to heavy grazing in sandy rangeland, Inner Mongolia*". In: Journal of Arid Environments 62 (2), p. 309–319.
Zha, Yong, and Jay Gao. "Characteristics of desertification and its rehabilitation in China." *Journal of Arid Environments* 37, no. 3 (1997): 419–32. doi:10.1006/jare.1997.0290.
[629] Jiang, "The Ordos Plateau", 30.
[630] Ibid.:30

abetted in the natural replacement of forage vegetation by non-forgeable plants, which could still grow despite the deteriorated soil.[631] This change in cover vegetation is an excellent example for certain rangelands in Inner Mongolia, like the Mu Us Sandy Land, which was subject to heavy grazing in the 1960s and 1970s.

Today, severe desertification in Inner Mongolia is also associated with the cultivation of unsuitable dry farmland[632]. Many scientists believe that the impact of erosion on croplands is seventy times larger than on pastureland, indicating that cultivation may have a greater effect on degradation than pastoralism[633]. Because of seasonal cultivation, which exposes the croplands in winter to storms, the soil cover is blown away. Eroded soils, in combination with the lack of precipitation and expansion of agricultural land, are to a large degree responsible for Inner Mongolia's desertification. The agro-pastoral transition zone is already covered with sandy sediments, showing a high degree of soil erosion. In combination with the naturally occurring heavy storms during the dry season the soil cover is in a constant and steep decline[634].

Although general climate changes and their human influence have impacted Inner Mongolia's environment, problems related to irrigation and water shortage as well as soil erosion and desertification can be connected directly to the population pressure, political reforms and the unsustainable handling of natural resources since 1950. The accelerated expansion of deserts from 1950 to 1990 and the increased human activity should both be seen as the primary cause for desertification.[635] Beside population increase and high animal stocking rates, major causes of human-induced desertification in Inner Mongolia are natural resource exploitation and unreasonable land-use practices. Before 1980, the land development policy, which focused on grain production, was responsible for large-scale field conversion. This led to most of the documented desert expansion between 1960 and 1980, followed by a steady recovery from mid-1980 to 2000.[636] The twenty years of continuous desertification increase in Inner Mongolia can be

---

631 Chen and Tang, "Desertification in north China", 371.
632 Ellis, "Sustainability of Inner Mongolian Grasslands", 176.
633 Jiang, "The Ordos Plateau", 68.
634 Wu, Bo, and Long J. Ci. "Landscape change and desertification development in the Mu Us Sandland, Northern China." *Journal of Arid Environments* 50, no. 3 (2002): 429–44. Accessed June 11, 2013.
635 Li, Hongwei, and Xiaoping Yang. "Temperate dryland vegetation changes under a warming climate and strong human intervention – with a particular reference to the district Xilin Gol, Inner Mongolia, China." *Catena* 119 (2014): 9–20.
636 Wang, "The relative Role", 54–55.

linked to governmental policies, which at the time were promoting increased livestock production and land reclamation, which resulted in overgrazing and over-reclamation. The vegetation recovery reported between 1980 and 1990 should mainly be attributed to afforestation, the reduction of arable land reclamation, grassland enclosure and a decrease in livestock production, which raises the question, whether human activity, despite the fact that is has continuously increased since the 1980s in this region, may also be capable of having a positive effect on environmental rehabilitation. From the 1950s to the 1960s afforestation increased but then constantly dropped until the mid-1970s.[637] Due to air seeding by natural wind distribution and greater investments in afforestation programs, the area of afforestation has steadily increased.[638] Nevertheless the recovered areas are still much smaller than cultivated areas, sandy land, grassland and deserts in the region, and therefore the data provided by the Chinese government, would suggest that the increasing areas of afforestation should be handled with caution, in terms of their credibility.[639]

To manage the desert expansion and grassland degradation in north China, the government released guidelines for herders. In the mid-1990s, as part of China's Grassland Contract Policy, pastureland was divided into small units for each household and the patches were fenced in in order to prevent trespassing.[640] Grassland enclosure, or "fencing-in", has in some areas improved the grassland quality and helped to rehabilitate vegetation. But as Dee Mack Williams already argued in 1996, the fencing-in of pastures and the political implementation of this policy failed to establish a sustainable system for pastoral production.[641] Chaotic and uncontrolled grazing practices outside the fenced in areas contributed to a high degree of desertification, especially on unenclosed public pastureland.[642] Since 2001, environmental conservation programs have been carried out that emphasize the rehabilitation of vegetation in order to prevent sand storms affecting Beijing and coastal urban agglomerations. These conservation programs have also contributed to a decline in livestock rates, as they utilize grazing pauses in spring, grazing bans in fragile areas, and reconverting

---

637   Wu, "Landscape Change", 440.
638   Ibid.:440
639   Ibid.:55
640   Wenjun Li, Lynn Huntsinger. "China's Grassland Contract Policy and its Impacts on Herder Ability to Benefit in Inner Mongolia: Tragic Feedbacks." *Ecology and Society* 16, no. 2 (2011): 1–14. Accessed June 8, 2015.
641   Williams, "Grassland enclosures", 307–314.
642   Ibid.:309

farmland to grassland.⁶⁴³ The recorded environmental recoveries correlate to the changing political focus on environmental issues, which became the subject of several governmental policies implemented after 2000.

In Inner Mongolia some areas like the Mu Us Sandy Land, Horqin Sandy Land, Hulunbuir Sandy Land and the Tengger Desert have all undergone recovery through dune fixation and re-forestation since the 1950s. Despite these early and on-going attempts, little success has been achieved until today. Only a few areas like the Mu Us Sandy Land have shown a stagnation in desertification since the 1980s, whereas the majority of Inner Mongolia's drylands have deteriorated to a tremendous degree.⁶⁴⁴ This will be discussed in the following section.

Northern China Dryland Expansion

To clarify, desertification is a geological process related to the expansion of drylands and deserts. Drylands are areas with sparse precipitation that are categorized in four subtypes: hyper-arid, arid, semi-arid, and dry sub-humid areas.⁶⁴⁵ More than 40 percent of the world's surface is covered with drylands and inhabited by 35 percent of the world's population living in poverty.⁶⁴⁶ The fragility of drylands lies in the dry overall climate resulting in poor soil quality with little water conservation capacity, which in turn leads to a low nutritional content in food grown in these areas. The vulnerability and sensitivity of dryland-ecosystems is then further exposed to climate change and human activities. Thus, clearly desertification affects drylands rather than humid or sub-humid regions.

Desertification and the expansion of drylands is a complex and diverse environmental issue induced by various natural and man-made factors. They can either be physiographic (climate, water deficiency, soil conditions and vegetation) or anthropogenic (intensive exploitation, increasing population, overgrazing, over-cultivation, deforestation, waste of water). Most of these environmental problems have already been discussed in previous chapters. Inner Mongolia has been confronted with all of these factors, which contributed to the dryland

---

643   Piao, Shilong. "NDVI-indicated decline in desertification in China in the past two decades." *Geophysical Research Letters* 32, no. 6 (2005). doi:10.1029/2004GL021764.
644   Ibid.:368
645   Li, Yue, Jianping Huang, Mingxia Ji, and Jinjiang Ran. "Dryland expansion in northern China from 1948 to 2008." *Advances in Atmospheric Sciences* 32, no. 6 (2015): 870–76. Accessed July 2, 2015.
646   "Global Drylands: A UN syste-wide response." Unpublished manuscript. http://www.unccd.int/Lists/SiteDocumentLibrary/Publications/Global_Drylands_Full_Report.pdf.

expansion and severe desertification in some areas. The dryland expansion since 1948 in north China can clearly be seen in the region. (Map 8) It is interesting that the dryland border in north-west China (Xinjiang, Qinghai, Tibet and Gansu) has been stable over the last seventy years, whereas the dryland border in the northeast of China, mainly along the middle-to-lower reaches of the Yellow River has definitely expanded further east and southwards. This correlates with the long-term trend of precipitation and evapotranspiration under a warming climate, which the region experienced from 1948 to 2008, as well as the expansion of semi-arid regions, as Li and Huang pointed out.[647]

The drylands in the northeast expanded by about 2 degrees longitude since 1994 and currently cover the whole of Inner Mongolia. The two red patches in southern Heilongjiang province turned into drylands since 2003.[648] In the south the drylands have expanded approximately 1 to 2 degrees latitudinally since 1994.[649]

Map 8: Expansion of Northern Chinese Drylands 1948–2008

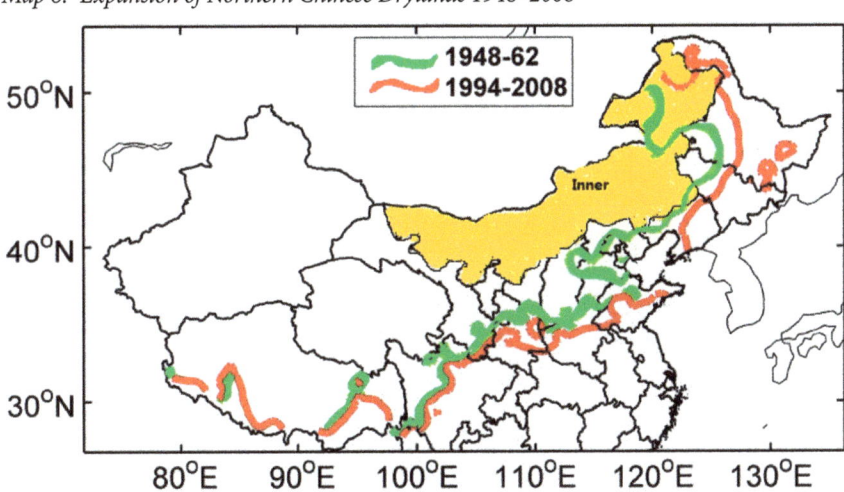

Source: Li, Yue; Huang, Jianping; Ji, Mingxia; Ran, Jinjiang 2015: "*Dryland expansion in northern China from 1948 to 2008*". In: *Adv. Atmos. Sci.* 32 (6), modified by Cilia Neumann (2016)

The increase of desertification in the same areas can thus be explained as additional pressure on the ecosystem. The expansion of desertificated land between

---

647 Li, Yue "Dryland expansion", 875.
648 Ibid.:873
649 Ibid.:875

1960 and 1980 correlates with the physiographic pressure mentioned by Li Yue and Huang Jianping (et al.) and the anthropogenic induced stress on the fragile ecosystem.

The following map illustrates the distribution of surface vegetation cover in north China and clearly indicates that the majority of Inner Mongolian landscapes are dominated by grassland, which is a typical surface vegetation distributed in drylands. The vegetation cover from dense grassland, steppe and semi-steppe to desert varies with the drylands. The fertile and dense "typical green" grassland can only be found in the northeast of Inner Mongolia, where precipitation is higher and a mesic[650] habitat is dominant. The second largest surface area is barren land, mainly in Xinjiang, Tibet (Xizang), Qinghai and Gansu. The distribution of crop-cultivated areas is comparatively small to grasslands. But included in the grasslands are the agricultural areas of pastureland. Having this in mind, this region takes on a whole new relevance: beside barren land most of the northern Chinese drylands are in use for agriculture, which as mentioned before increases the degradation of the drylands.

Map 9: Vegetation Cover in North China

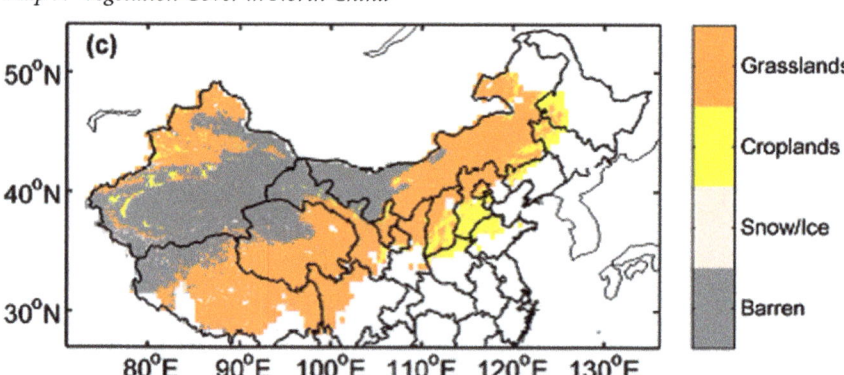

Source: Li, Yue; Huang, Jianping; Ji, Mingxia; Ran, Jinjiang 2015: "*Dryland expansion in northern China from 1948 to 2008*". In: *Adv. Atmos. Sci.* 32 (6)

Historically it has been shown that in Inner Mongolia the estimated boundary shift of potential suitable agricultural area has always moved directionally with

---

650   *Mesic* is one of the terms to describe the amount of water in a habitat: hydric, mesic and xeric. A mesic habitat is a habitat, where the water or moisture supply is well-balanced or moderate.

temperature changes and changes in annual precipitation. Altogether the boundary shift has moved within an approximate range of 75 to 250 kilometers southwards.[651] Today, Inner Mongolia is marked by "fixed" boundaries of agricultural areas and pastures, so shifting according to climate changes is no longer possible. Therefore the national strategy for agricultural land planning and pasture management must include the natural environment and question whether the climatic conditions are appropriate for crop cultivation or livestock production. Governmental policies should emphasize the rehabilitation and protection of climatically fragile regions. The GfG policy which returns cultivated land into grassland and replaces cropland with pastures could be a successful strategy in helping to restore the deteriorated environment.

Being affected by environmental distress, the increase of desertification or the expansion of drylands are all major problems for farmers and herdsmen in Inner Mongolia. Through the accumulation of sandstorms and expansion of moving dunes, the burying of large areas and whole villages under mountains of sand is a legitimate threat to the local population. Climate changes and societal factors, as mentioned previously, have helped to shape Inner Mongolian landscapes and contributed to the environmental problems occurring today. The exploitation of Inner Mongolia's farming-grazing transitional zone has also had a significant impact on the drylands. Additionally, water shortages and sinking ground-water tables due to unsustainable water use, overgrazing, land reclamation and all the other man-made interferences have further contributed to the severe problems. Thus the large-scale desertification of north China cannot and should not be connected to a single event or cause, as several scholars have pointed out. The author fully agrees with the opinion, that adverse natural conditions in combination with human activities are responsible for environmental degradation.

Managing the progressive loss of soil and grassland as well as limiting the damage of further land degradation is of high priority for the national government and the local people in Inner Mongolia, which will be the topic of the following chapter. A few national and international aid projects to narrow down desertification in Inner Mongolia will be presented.

---

651 Ye, Yu, and Xiuqi Fang. "Boundary shift of potential suitable agricultural area in farming-grazing transitional zone in Northeastern China under background of climate change during 20th century." *Chinese Geographical Science* 23, no. 6 (2013): 655–65.

7.3 Dealing with Desertification: National Policies versus NGO Based Action Programs in Inner Mongolia

There are several projects combating desertification in Inner Mongolia. Some are supported by the Chinese government and concentrate on desertification issues nationwide with a particular focus on the northern Chinese desert and dryland belt.[652] Other small scale projects have been initiated by non-governmental organizations and are working at on-site locations. This section will illustrate the Chinese government's efforts to stop desertification in cooperation with the Food and Agriculture Organization of the United States (FAO) as well as some of the smaller independent projects based in Inner Mongolia. The projects' common objective is the reforestation of desertificated regions in order to combat dryland expansion. To reach this goal, the Chinese government and non-governmental institutions have been planting trees and green shelterbelts to stop the dunes from expanding.

National Measures and Policies to Combat Desertification in China

China's encouragement in the development of forestry, improving the environment and combating desertification dates back to the 1970s when the *Three North Forest Shelterbelt Program* (*Sānběi Fánghùlín* 三北防护林) was first carried out.[653] According to the Chinese government's master plan, the program will be completed in 2050. The program is based on three initial stages (1978 to 2000, 2001 to 2020 and 2021 to 2050). The goals of the first period have been accomplished, including the shelterbelt construction for farmlands, natural forest protection, restoring farmland to forest and the controlling of sand dunes around Beijing and Tianjin. Approximately 590 counties (as of 2010) are or have been involved in the project as well as thirteen provinces in north-west, north and northeast China.[654] The project is not only supported from within China but also by organizations outside of the country.

---

652 NGO initiated projects are the "Green Belt", the "Timberland© Horqin Reforestation Project" and the "Greening Earth Project". The Chinese government has carried out reforestation projects since the early 1970's (http://www.fao.org/docrep/003/y1237e/y1237e10.htm#TopOfPage).

653 Wang, X. M., C. X. Zhang, E. Hasi, and Z. B. Dong. "Has the Three Norths Forest Shelterbelt Program solved the desertification and dust storm problems in arid and semiarid China?" *Journal of Arid Environments* 74, no. 1 (2010): 13–22. Accessed September 20, 2015.

654 Ibid.:14

The United Nations Convention to Combat Desertification (UNCCD) has been encouraging China in this project for several years and its influence has been both direct and indirect in helping to raise public awareness on the issue as well as promoting China's desertification control activities and adopting advanced technologies. The UNCCD is additionally involved in strengthening desertification monitoring and warning systems, supporting China in establishing a legal system for desertification prevention and control as well as helping China to implement attractive policies and mechanisms to invest in forestry and desertification control. The UNCCD helps to implement National Action Programs, which are based on the *Chinese Agenda 21* and the Three North Program. The program has been financed through various channels: central government allocations, locally raised funds, interest loans and overseas financial support, together with money provided by the Ministry of Finance.[655] The National Action Program intends to control and combat desertification in three stages.

1. By the year 2000: The continuous expansion of deserts should have been mitigated, the environmental conditions improved and the living standards of the people improved.
2. By 2010: The aim was a considerable improvement in desertification-affected areas and living standards.
3. By 2050: The desertificated areas will be effectively controlled and the total area of nature reserves should cover more than 90 million hectares. The economic and environmental conditions should be on-level with national averages for the desertificated areas.[656]

So far, improvements in desertification control and technology adoption have been quite successful in the Shulinzhao Township in Dalate Qi in Inner Mongolia. In 1996, together with the cooperation of the National Bureau to Combat Desertification (NBCD) in China, the UNCCD was able to establish Shulinzhao as a demonstration site to combat desertification on community level. Now twenty years later, Shulinzhao shows significant forest and woodland recovery.[657] Shulinzhaos' success was based on the adoption of

> Many new technologies, such as water-saving irrigation, drip irrigation, use of greenhouses, use of a soil conditioner that improves the soil's water retention and enhances

---

655 Liu, Tuo. "Influence of the Convention to Combat Desertification on forestry in China." In *GLOBAL CONVENTIONS RELATED TO FORESTS*. Edited by A. Perlis. Unasylva 206, 2001. http://www.fao.org/docrep/003/y1237e/y1237e10.htm#P0_0.
656 Ibid.
657 Ibid.

fertilizer availability, use of plastic mulching material and advanced rice cultivation techniques.[658]

An interesting fact about the Shulinzhao model was that due to poverty and food shortages, most of the men had to migrate to find work and mostly older people, women, and children remained in the village. The villages' success referred to the organization of the Women's Union of the Township, which organized the fight against the sand dunes.

But the government has also employed other methods with the goal to raise environmental awareness and to mobilize the public. Since the first World Day to combat desertification and drought in 1995, China has actively supported large-scale public awareness raising activities, such as organizing groups for tree planting in commemorative forests (women, People's Liberation Army, Communist Youth League). Since then every year large numbers of volunteers, including the old, children, workers, farmers, officials and soldiers take part in these activities.[659]

In addition to the engagement on local levels, the Chinese government has also been working to fight the desertification issue on national level. With the implementation of the China Coordination Implementation Committee to Combat Desertification (CCICCD) the Chinese government has been able to bring together several sectors that combat desertification and contribute to UN-CCD. The railway and communication sectors are working in support of the implementation of green corridor programs, the control of soil and erosion in desertification-affected areas and the regulation of the water resources associated with that sector. The agricultural sector has contributed to the development of an eco-agriculture in China. Beyond this, the People's Bank of China has increased discounted loans with an annual contribution of $140 million for programs to combat the expansion of deserts and drylands.[660]

Additional steps that the Chinese state has taken include the dissemination of advanced knowledge and supporting the distribution of innovative technology to combat desertification. Technical service is provided through R&D institutions. The CCICCD in particular has taken on the job of implementing thematically-focused trainings and information on the issue of desertification. Since 1996, the CCICCD has been actively supporting the people in combating desert expansion.[661] In 1996, the CCICCD started to implement the Law of Combating

---

658 Ibid.
659 Ibid.:5
660 Ibid.:5
661 Ibid.:5

Desertification, which helps to guide improvements being put forth within the legal system. The law's specific focus was to establish a sustainable development strategy that would prevent areas from further desertification, bring it desertification under control generally, and manage it more sustainably. The Chinese government has also promulgated more than twenty laws, by-laws, and regulations related to environmental protection. Governmental efforts lie more in guaranteeing economic and social developments that encourage ecological and environmental improvement on the basis of grassland and forest protection.[662]

To encourage and engage people to support afforestation projects and programs to combat desertification, the Chinese government has adopted several different measures and policies in the last twenty years. These include discounted loans for anti-desertification projects as well as tax deduction or tax abatement for the development of barren hills and sandy drylands.[663] A result of this has been that more businesses are engaged in combating desertification and have turned to ecological industries, which support afforestation and shelterbelt programs in north China.

Finally, the United Nations Convention to Combat Desertification (UNCCD) has promoted international cooperation for desertification issues since the 1990s. Within an international cooperation and partnership with Germany, China has established reforestation programs from 1993 to 1996 and the German government has assisted in desertification control and ecological rehabilitation projects in Inner Mongolia and Liaoning province.[664] Since the UNCCDs implementation in 1996, Germany hosts the Secretariat of the UNCCD in Bonn.[665] Still more than 10 years after the UNCCD came in operation, Germany views the organization as a key instrument for its strategic development cooperation. In 2005 Germany funded more than 679 projects worldwide with an overall investment of 1.8 Billion EUR, implementing the UNCCD's approach in combating desertification, supported by various state and non-state organizations.

---

662 Tao, Wang, and Wu Wei. "Combating desertification in China." *Engineering construction* 379 (1998): 50–64.
663 Liu, "Influence of the Convention to Combat desertification", 7.
664 Ibid:8.
665 UN. "Desertification and Drought: Strategic Framework for combating desertification in Germany's development cooperation: UNCCD, MDGs and the Paris Declaration." http://www.un.org/esa/agenda21/natlinfo/countr/germany/desertification.pdf.

Non-governmental Programs Combating Desertification in Inner Mongolia

Numerous non-governmental organizations (NGOs) from around the globe have also been involved in combating desertification from the Sahel zone in Africa to the desert belt in Asia. As mentioned previously, the UNCCD has been active nationwide in encouraging the country's efforts to combat dryland expansion. In combination with NGOs, the Chinese government has reported success in environmental protection projects related to dealing with dryland expansion in Inner Mongolia.

To present a broad overview of NGO-based environmental protection actions, three representative non-governmental organizations have been selected. These actions are engaged in different areas in Inner Mongolia supporting reforestation projects and appealing for ecological awareness through the involvement of the local population in their environmental protection programs. These three NGO-based projects are the Greening Earth Project[666], established in the 1990s, the Green Belt Project[667] implemented in 2001, and the Timberland© Horqin Desert Reforestation Project, initiated in 2010. These programs are all based on the method of afforestation.

Greening the Earth is a project, which was launched under supervision of the greening company "Green Planet", a Japanese non-profit organization (NPO) established in 1994. Its aim is to combat desertification in Inner Mongolia.[668] Since 1997 the organization worked on site in Baierye Dune, near Hohhot City, starting with the conversion of 2000 hectares of desert dunes and fields into bushes and trees. A second project was launched in 2006 in the desert of Kubuqi, Dalate Prefecture near Ordos City. With the establishment of the "Greening Trust" the organization built up a participating system via donation for everybody, who wanted to take part in the project. The participator can choose a wide scale of donation possibilities, according to the greening area from 50 square meters (12.5 U.S. Dollars) up to 10.000 square meters (2.500 U.S. Dollars). The greening funds are only limited to project related purposes, like purchasing plants, equipment and machinery as well as local workers labor expenses. Local residents and

---

666 Greening the Earth is available on: http://www.ecostyle.net/.
667 OISCA International available on: http://www.oisca-international.org/programs/environmental-conservation-program/china/establishment-of-green-belt-to-combat-desertification-project-in-inner-mongolia-china/.
668 Tatsushi Masuda. "Twenty-Year History of an Activity to Prevent Desertification in Inner-Mongolia, China: 内モンゴル沙漠化防止活動の20年." *Journal of Group Dynamics*, 2014, 3–71; 集団力学集団力学. doi:10.11245/jgd.31.3.

Greening Trust Agents are on site responsible for the implementation of the project. Donation is possible via the homepage of the ecostyle.net, which was initiated in 2000, serving as a platform for ecological activities. Until 2010 this NPO has further invested in R&D, participated at the "World Summit on Sustainable Development in 2002" in Johannesburg, South Africa and established an "Inner Mongolia Green Friendship Center" in Baierye Dune in 2006. In 2009 and 2010 the organization had held the first and second "Inner Mongolia Summer College for Environmental Study" at the Inner Mongolia Green Friendship Center, where students from Japan, China and the United States have participated.[669]

In 2014 Tatsushi Masuda of NGO ecostyle.net retrospectively took a stock of the last twenty years being engaged in desertification projects in Inner Mongolia.[670] After analyzing and discussing the organizations approach of developing an international network among interested parties, including organizations, which support environmental conservation programs as well as local communities promoting and participating in environmental activities, Masada's analysis is disillusioning. The conclusion drawn is, that the activity in Inner Mongolia has changed over the past twenty years. He explains that the project failed during the initial state (1995–1999), when the project attempted to combat desertification with agricultural development. Then the focus was changed to community activities and the establishment of a sustainable agricultural circle within the village, with emphasis on renewable energy supply and a pasture-livestock balance. Although the project has reached participation outside the organization and established a large network of people, parties and groups interacting with each other, the project stagnates since 2008. Here Masuda refers to the most interesting aspect of his analysis. Beside shortage of financial resources, the main problem is the local population's lack of environmental attitude, which urges them to get engaged in pro-environmental activities. A point, which was confirmed during field researches in Inner Mongolia by the author himself, which will be represented and analyzed in chapter 8.

The Green Belt Project was developed by the Organization for Industrial, Spiritual and Cultural Advancement-International (OISCA), which is based in Tokyo and strives to contribute to environmentally sustainable development connected to "agriculture, ecological integrity and human spirit."[671] OISCA's approach has been to involve locals in activities, advocate hands-on experimental

---

669 http://www.ecostyle.net/e/aboutus.html.
670 Tatsushi Masuda, "Twenty-Year History of an Activity to Prevent Desertification in Inner-Mongolia", 71.
671 http://www.oisca-international.org/about/mission/.

programs, transfer knowledge and skills, and cultivate awareness by supporting a "spiritual" community of dedication, self-reliance and ecological integrity.

Within the Green Belt Project, OISCA was able to mobilize Inner Mongolian locals and Japanese volunteers to plant trees, covering an area of 800 kilometers by 14 kilometers in order to establish a green belt in Alashan. OISCA also set up an ecology institute in Alashan, where R&D programs to measure desertification are conducted. Innovative techniques like seed-spreading with airplanes are also methods that have been utilized to support afforestation in Alashan.[672] The Japanese company Brother Industries (printers, multifunction fax machines etc. pp) is also engaged in reforestation projects in Inner Mongolia, in cooperation with OISCA.[673]

The Timberland© Horqin Reforestation Project was initiated by the outdoor company of the same name.[674] Timberland is already involved in several reforestation projects in Haiti and China, and supports events and organizes employee-led environmental projects within its operating community. Horqin, situated on an alluvial plain was previously green grassland. It is now the Horqin desert and covers an area of approximately 42,300 square kilometers, or nearly the size of Switzerland.[675] Every year the desertification of Horqin is expanding at a rate of roughly 10,000 square kilometers. In Horqin, overgrazing and climate changes have contributed to the massive desertification, and one consequence of this process, are the sand storms caused by the westerly wind drift over the exposed surface soils. These sand storms in Horqin have impacted the air traffic in China, Japan, South Korea and Taiwan. To help combat these issues, the Horqin Projects has planted 1,653,670 trees in the Alashan region since 2001.

Timberland© is also engaged in the education of the local population, establishing communities and training its own staff in raising awareness for environmental issues. One of the signs that this project has been successful is the fact that many families, which had previously left Horqin because of the severe desertification, have now returned to their homeland. With the additional government-supported incentives many locals also actively support the reforestation project.[676] Over the last fourteen years Timberland© has invested roughly 6 million RMB and 291 days of voluntary employee time in its reforestation

---

672 Ibid.
673 http://www.brotherearth.com/en/top.html.
674 Horqin Reforestation Project available under: http://www.timberland.com.sg/horqin/our-mission/.
675 Ibid.:2
676 Ibid.:3

project. Timberland© also works in cooperation with the Japanese NPO Green Network (緑化ネットワーク), which educates the local government and Inner Mongolian communities on desertification issues in the Horqin Desert since 2001. Together they have been working towards forest maintenance and woodland conservation in north China.[677]

To conclude the engagement of NGO's in Inner Mongolia, one aspect of specific interest should be mentioned here. It astonishes, that the majority of projects carried out in Inner Mongolia is initiated by Japanese environmentalists and organizations.[678] This engagement dates back to 1998 after the Yangtze flooding, when the two countries agreed to long-term cooperation in environmental conservancy and reforestation.[679]

---

677 Green Network: "緑化ネットワークは、中国ホルチン砂漠だけではなく、国内でも森林整備や里山保全活動を行っています". Available under: http://www.green-network.org/gn/index.php?option=com_content&view=article&id=164&Itemid=54.
678 Barthélémy Courmont. "Towards a "Green Detente" between Japan and China? The Case of Cooperation on Reforestation." *Issues & Studies* 51, no. 3 (2015): 29–62.
679 Information available from the Japanese Business Federation on: https://www.keidanren.or.jp/english/policy/2001/006.html#part2.

8. Empirical Research II

> *"Problems cannot be solved at the same level of awareness that created them."*
>
> Albert Einstein (1869–1955), Theoretical Physicist

This chapter focuses on the development of environmental awareness in Inner Mongolia as it relates to environmental concerns. This is an analysis of the interviews conducted during field research in 2013. The results have been correlated to the development theory of environmental awareness and the work of Hans-Joachim Fietkau, Hans Kessel and James Blake with the aim to answer the questions: How do people in Inner Mongolia respond to environmental issues?; How do they face these issues?; and, Does knowledge of these environmental concerns urge them to act pro-environmentally?

8.1 Developing Environmental Awareness and Pro-Environmental Behaviour in Inner Mongolia

> *"A society is defined not only by what it creates, but by what it refuses to destroy"*
>
> John C. Sawhill 1936–2000, CEO of the Nature Conservancy

There is a clear gap between having environmental knowledge and awareness versus adopting pro-environmental behaviour[680]. With reference to chapter 1.4.3, the approach of this chapter is to explain the "value-action gap"—or the lack of action that often occurs despite knowledge of environmental problems—between environmental attitude and behaviour in Inner Mongolia. The question is: Why is it that even when people are aware of environmental problems they rarely develop pro-environmental behaviour? The research conducted in Inner Mongolia revealed that only half of the interviewees showed little or no engagement in pro-environmental action, unless the people are not involved in NGO projects or governmental mass mobilization programs, and also do not benefit from financial subsidies.[681]

---

680 By pro-environmental behaviour the author understands the individuals' awareness of minimizing destructive effects on the natural environment. Minimizing resource-extraction and reduction of pollutants and waste as well as sustained actions (reforestation) are pro-environmental behaviours.

681 Detailed analysis of the interviews will be discussed in this chapter and portrayed under 8.2 (Case studies).

As mentioned in the previous chapter, establishing awareness and pro-environmental behaviour on the local level are both important components of the work of NGO-led environmental protection programs. To achieve sustainability within these programs both pro-environmental awareness and action are required. In Inner Mongolia, three institutions have been involved in the process of developing environmental sustainability and awareness as well as trying to engage the local population in pro-environmental behaviour. The following section describes the interaction and influence of these three participants in the process of raising environmental awareness. The responsibilities of NGOs, the government and local populations will be outlined in order to illustrate each of their roles in this process of awareness raising and pro-environmental behaviours.

NGOs are interest groups that function as the link between local government and community. In Inner Mongolia they have established cooperative programs, which engage the community and government together with that particular organization. Development programs, educational trainings and environmental action projects have also been initiated or supported by NGOs active in the region. Reforestation projects, like establishing green shelterbelts have been implemented with the help of volunteers, recruited both from within and outside of the NGOs. These NGOs that are active in the environmental issues of Inner Mongolia typically have the shared main objective to raise the people's awareness and present solutions through on-site educational training and knowledge transfer. The NGOs active in the region like Timberland© or Green Network hold an influence mostly on the local level, where the interaction with the resident population or local government is the main interest.

Fig. 6: Environmental Awareness and Action Building Process in Inner Mongolia

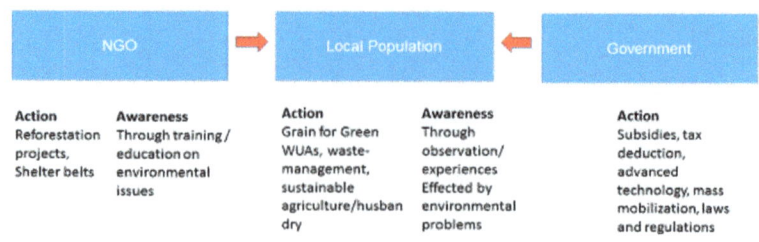

Source: Cilia Neumann (2016)

The government's role in the environmental protection process has since 2011 been based on the guidelines of the Twelfth Five-Year Plan, which in contrast to previous plans places a larger emphasis on the environment. Therefore the development of environmental protection and sustainability efforts refers to the implementation of laws and regulations with the explicit purpose to move the local population to pro-environmental actions. The national government's responsibility has been to guarantee the realization of these environmental guidelines and to maintain a level of order in their implementation. With the Twelfth Five-Year Plan raising awareness through educational programs has mainly been replaced by financial subsidies, which encourage the producer or consumer to act pro-environmentally. The Grain for Green program is representative, creating incentives for afforestation projects because its desired outcome is to increase the amount of reforested areas and shelterbelts as a method to measure pro-environmental behaviour.

The local communities in Inner Mongolia are the main actors in the process of developing environmental awareness and action. They benefit from financial support from the national government as well as educational training and knowledge transfer, which is provided by engaged non-profit organizations. In the best case scenario, the local population is embedded in the development stages of both governmental benefits and non-governmental activity programs. When this happens they have the possibility to develop awareness and attitudes from independent sources, a process which has suggested an ability to help close the value-action gap associated with pro-environmental behaviour in Inner Mongolia.

8.2 Case Study 3

The following section will assist in revealing the difficulties of applying theory to practice and help to explain how pro-environmental behaviour can develop through the attainment of environmental awareness in Inner Mongolia. In this situation, it is clear that the value-action gap is similar to the idea of a "theory-practice gap," meaning the gap that exists between supporting pro-environmental actions in theory, and not actually following through with these beliefs in practice.

Applying environmental theories like Kuznets Curve or those approaches for developing awareness would seem to be applicable in a much wider context than Kuznets proposes. For example, some research has illustrated that Kuznets assumption works for certain pollutants, like carbon dioxide emissions.[682] Since the

---

682 Susmita Dasgupta, Benoit Laplante, Hua Wang and David Wheeler. "Confronting the Environmental Kuznets Curve." *Journal of Economic Perspectvives* 16, 147–168 (2002). Accessed December 8, 2014.

introduction of the Eleventh Five-Year Plan, China has continuously reduced its emissions.[683] Still, the reality of the situation is that industrialized countries like China typically show a decline in some pollutants with economic growth, but it is often the case that as an improvement in one area leads to a negative change in another. For example, other environmental issues may increase, such as water-shortage and land erosion in Inner Mongolia. Thus, the theoretical hypothesis that environmental degradation decreases with rising per capita income is more complex and environmental issues do not diminish, but instead are shifted to other problems. Thus is it correct to assume that an increase in living standards, which includes the growth of average income levels, security and access to education lead to environmental awareness? Generally speaking, yes, environmental knowledge might increase overall environmental awareness, but individual situations and the willingness to change the individual behaviour remains more difficult to measure with certainty.

When applying the linear progression approaches of Fishbein and Ajzen, Rajecki and Hines, Hungerford and Tomera, both of which assume that the key to pro-environmental behaviour lies in the availability of education and knowledge, the results of the field research in Inner Mongolia only partially confirms these theories. Although the majority of the people in Inner Mongolia do have access to education beyond the elementary level, it is difficult to start from the premise that their educational background has been the major or main influence on their environmental behaviour. As the diversity of answers illustrate, it is not only environmental knowledge, which seems to stimulate the people's environmental awareness and pro-environmental action, but also the observation and experience of environmental problems. Thus, the only consensus in line with these theories is that of all people interviewed only three have not attended school and demonstrated no pro-environmental behaviour. Although they did show established sense of awareness in terms of the root causes responsible for environmental degradation, this result is essentially negligible. Therefore, the linear theories of Fishbein and Ajzen, Rajecki, Hines, Hungerford and Tomera are not sufficient to explain the Inner Mongolian value-action gap. Diverse factors influence pro-environmental behaviour, such as those suggested by Fietkau and Kessel and Blake, who take into account that external or internal sociological as well as psychological factors play a major role in the development of

---

683   Jalil, Abdul, and Syed F. Mahmud. "Environment Kuznets curve for CO2 emissions: A cointegration analysis for China." *Energy Policy* 37, no. 12 (2009): 5167–72. Accessed November 8, 2015.

pro-environmental behaviours. For Inner Mongolia, with the specific concentration on Siziwang as a generic smallholder's potato production site, external factors (economic, sociological and political constraints, which might hinder the person to act pro-environmentally) are the dominant determinants that influence a person's decision. Blake's "Three Barriers" (Individuality, Responsibility and Practicability) are the second largest obstacle to pro-environmental behaviour. The people in Inner Mongolia are not *in principle* averse to environmental issues and are thus receptive to proposals and solutions to reduce the environmental damage. Still, the results of the surveys demonstrated that only 55 percent of the people are engaged in pro-environmental behaviour. A lack of interest, information and confidence in individual initiatives was a frequently cited reason to defend inaction. The questioning efficiency, facilities and costly activities were also shown to be further barriers for the people in Inner Mongolia to get involved in pro-environmental behaviour. These combined results thus confirm that the current environmental situation in Inner Mongolia calls for immediate action from the government to close the huge awareness gap and the non-involvement of the local population.

8.2.1 Siziwang Potato Smallholders

The following analysis refers to the evaluation material provided by interviews with the potato smallholders in Siziwang Qi, Inner Mongolia.

Due to the topic's sensitivity this, the first objective of the interview was to gain the trust of the interviewees and develop their confidence in the questionnaire. The first two parts of the interview focused on the individual's family situation and professional life and were subject of earlier analysis. The farmers could freely provide personal information on age, education and family members, which in turn helped to establish a better understanding of their perspective on the future. Before turning to the survey analysis it should be acknowledged that the situation of farmers in Siziwang should not be compared to that of farmers in western countries such as Germany. This is due to multiple reasons. First, the majority of farmers in Siziwang are smallholders operating at a subsistence level with the ability to sell a very small portion of their production on the local market. In Germany, farmers are, generally speaking, a trained professional whose main occupation is agriculture and at least 50 percent of his income is derived from it. The fact that individual land rights do not exist in China is the second aspect, which differs greatly from common western interpretations of ownership. The idea of the government seizing private property is something inconceivable to farmers in Germany. Therefore the concept that a farm should or could be managed and maintained over generations with "tradition" as a motivator is not embedded in the thinking of farmers in Siziwang. Keeping this in mind, it becomes understandable why living as a smallholder in Siziwang is not a matter of choice, but rather a pragmatically motivated decision to earn a living, even if barely above the subsistence level. Lastly, smallholders in Inner Mongolia are often excluded from access to markets, water resources and the transfer of new technologies which increase profitably and allow for the production of surplus products that can then be sold. Without government subsidies and support from various international agribusinesses which provide seeds, treatment and technological equipment as well as financial allowances, smallholders in Inner Mongolia have difficulties to compete economically. Farmers in Inner Mongolia are heavily subsidized to produce food, which would be cheaper if bought from abroad. The agricultural sector is therefore encouraged to expand on unsuitable land to cultivate staple. In 2012 the Chinese government has invested $165 billion in direct and indirect agricultural subsidies, followed by Japan, which has spent $65 billion and America over $30 billion, according to research by OECD.[684] Whereas the trend in the OECD countries goes to a reduction of governmental subsidies, China which has reached the point to hardly rely on its own agricultural production, invests massive costs to maintain on its self-reliance and shows an increase of its producer support estimate (PSE). The costs of agricultural subsidies will continue to rise, not least of labor, as the young generation migrates into the cities.[685]

The first question in the interview related to personal attitudes referred to the perspectives of the farmers for their future and retirement. An interesting fact that should be mentioned is that the farmers often explained that they do not want their children to receive the typical inheritance of a potato smallholder in Siziwang. The majority of the interviewees reported that they support their children financially, with the hope that they are able to receive a higher level of education than their parents. Many argued that an academic education is the key to financially secure jobs in big cities. Often the farmers also explained that their pension plan is to follow their children into the bigger cities and as a sort of *quid pro quo* be financially supported in their retirement by their children. This attitude appeared contemptuous in terms of keeping older traditions alive and

---

684 "China and global farming the wrong direction. As others cut farm support, China spends more", May 16th 2015. Available at: http://www.economist.com/news/china/21651272-others-cut-farm-support-china-spends-more-wrong-direction.
685 "Farm subsidies Bitter harvests. A drive for self-sufficiency in food comes at a growing cost", May 16th 2015. Available at: http://www.economist.com/news/china/21651276-drive-self-sufficiency-food-comes-growing-cost-bitter-harvest.

holding on to the family business as an achievement, which is often viewed as something worth preserving. But with the knowledge of these 'retirement plans', this decision to abandon family farms becomes somewhat clearer.

Against this background, answers regarding environmental awareness and pro-environmental behaviour are easier to understand. To clarify, the given answers do not explain whether the farmers' reasons for pro-environmental behaviour were politically or personally motivated. That is a question of interpretation, where the individual background is essential to interpreting the provided answers.

Environmental Awareness in Siziwang and the Difficulty of Developing Pro-Environmental Behaviour

The majority of Siziwang smallholders (80 percent) receive annual governmental subsidies between RMB 1,000 and RMB 7,000. All interviewees reported fearing environmental degradation, but only 50 percent actually developed environmental awareness in which they reflect on the reasons for the ecological damage. Looking at the number of people whose environmental awareness led to pro-environmental behaviour, it shrinks further to 40 percent within this category. The remaining 20 percent of farmers who receive no financial subsidies are also consequently making no attempt to act pro-environmentally, nor did they show deep environmental awareness for the environmental concerns (drought and soil erosion) affecting them. At the moment, it would be too narrowly considered to deduce a rule of great generality, but the coincidence might give space to the assumption that governmental subsidies could be a driver for pro-environmental behaviour.

Besides household subsidies, re-investment from agricultural net income is also used to achieve farming improvements with the assumption to positively influence environmental issues. Therefore most of the farmers (70 percent) reinvest in machines, seed and irrigation improvements, like plastic mulch or green houses as well as drip irrigation systems—all of which are deemed to be pro-environmental and water-saving irrigation technologies. Although water scarcity is their main environmental concern, only 10 percent of the interviewees have invested in drip irrigation systems or other irrigation equipment. But when it comes to the question of prevention methods, the same farmers stated that they do not currently do anything to prevent further soil erosion and drought issues on their farmland.

Only 33 percent reported trying to solve the water shortage with new wells to irrigate their fields. In the belief that this action prevents further environmental degradation, the farmers are absolutely convinced that this is a demonstration of pro-environmental behaviour. But as mentioned in chapter 7.2, the digging of these additional wells has increased Inner Mongolia's water scarcity as they cause the groundwater table to further drop. Hence, what seems to be personally motivated pro-environmental behaviour is in actuality contra-environmental. The remaining 13 percent of interviewees explained that they plant trees and use organic manure to fertilize the exhausted soils, which they believe are useful for the prevention of environmental damage.

In terms of irrigation issues, the Siziwang potato growers demonstrated awareness of the water shortage and predominantly claimed that this was due to naturally occurring conditions. Of the farmers questioned, 93 percent had viewed drought, soil erosion, salinization and alkalization as their main threats. However, their opinions as to the root causes differed from climate change, global warming, misuse of fertilizers and sandification. Only 10 percent of those, who mentioned salinization and alkalization as severe restrictions for their agriculture had tried to solve these specific problems through the application of additional sheep dung. This behaviour is thus reasonable when it is taken into account that these farmers believe that their soils are nutritionally depleted and therefore the appropriate solution is the application of additional soil nutrients. The suggestion that over fertilization may also have a negative impact did not seem to be of interest to the interviewees.

With specific regard to environmental attitudes and the coincidence of environmental awareness and pro-environmental behaviour the results suggested that the value-action gap in Inner Mongolia can mainly be attributed to disinterest. The majority—93.3 percent—fear environmental degradation, but only 53.3 percent of the interviewed smallholders are simultaneously reflecting upon the causes for these problems. Within this peer group only 40 percent of the total farmers, who perceive of environmental awareness and as a consequence show pro-environmental behaviour. The same amount of farmers demonstrated no environmental awareness and were consequently not engaged in any pro-environmental action. The rest of the peer group, merely 20 percent, were inhomogeneous in terms of their actions and behaviours, meaning that they are either aware of environmental problems and reflect upon their causes but make no attempt to act in an environmentally friendly way or they are realizing the obvious agricultural problem, which is drought, but they are not aware of the causal relation, but instead try to solve the problem with individual initiatives such as digging wells. Because these answers often reflect a certain indifference towards environmental issues, it could interpreted that there is overall little interest in environmental issues in this group.

Of the whole group interviewed, half demonstrated some level of environmental awareness and pro-environmental behaviour, while the other half did not. Ignoring the relatively low number of people (13 percent), who possess environmental awareness, but obviously doubt the value of acting in a pro-environmental manner, the interviewees can be divided thusly: 46.6 percent lack environmental awareness, while 53.3 percent do not operate their farms in a pro-environmental manner and therefore demonstrate a clear lack of interest in the prevention of further environmental prevention. A relatively insignificant group of 6.6 percent did not show any awareness or attitude towards environmental issues, but instead only a fear of drought. Nevertheless they were actively trying to prevent irrigation issues through the drilling of wells.

Education and Pro-Environmental Behaviour in Siziwang

Often the answers to explain individual inaction were: "I don't know", "I don't care" and "I didn't try". The frequency of these statements therefore raised the question of whether any relation or connection can be made between education levels and the development of environmental awareness within the group of potato smallholders. To provide a short insight in China's educational system, the following illustration has been included.[686]

Fig. 7: China's Education System

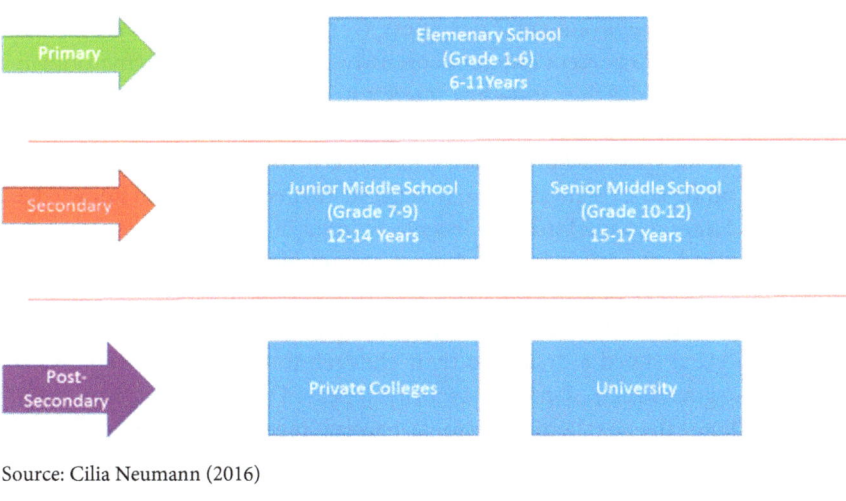

Source: Cilia Neumann (2016)

---

[686] For detailed information please see: http://www.classbase.com/Countries/china/Education-System.

In general, the potato farmers in Siziwang had a solid primary education. In fact, 60 percent graduated from junior or senior middle schools.[687]

Eleven of the farmers had only visited Elementary School, and nine of these farmers were not engaged in pro-environmental behaviours, representing more than 80 percent within this classification. Nevertheless, six of the farmers did demonstrate environmental awareness, in the way that they feared drought, soil erosion, salinization and alkalization as threats to their agriculture and they speculated about the reasons for these problems. Their answers were overall quite reflective and ranged from general climate changes, to human pressures and the misuse of fertilizers and pesticides. However, only two of six interviewees were trying to prevent a further increase of these issues. The farmer who reported feeling threatened by drought and a rise of warm climate also believed in additional irrigation through wells. The farmer, who was affected by salinization and blamed synthetic fertilizers for the problem, utilized organic manure to dung his field. The remaining four farmers had made no attempts to combat the problem and had not tried any prevention method.

Eighteen smallholders went to junior and senior middle school, but only one-third did not report acting in a pro-environmental manner. The same farmers also did not demonstrate any knowledge of environmental protection awareness or specific attitudes towards their environment. The results of the surveys did suggest, however that farmers with a higher level of education had typically developed some level of environmental awareness. Of the farmers with an elementary level of education, 83 percent did not demonstrate an awareness of environmental degradation nor did they show any interest in acting pro-environmentally. In contrast, those farmers with a somewhat higher education level, only 33 percent demonstrated the same lack of awareness or concern Thus the results of the group of farmers surveyed in Siziwang clearly suggests that education and knowledge, as Rajecki, Hines, Hungerford and Tomera pointed out in their linear progression approach, is the basis for developing environmental awareness and pro-environmental behaviour.

When asked about their hopes for their children's education and future prospects, 50 percent stated a desire for their children to become civil servants or work for the government. Financial security appeared to be an important factor in this thinking. The next largest group, 43 percent, reported wanting to see their

---

687 As previously mentioned, only one farmer reported not having attended school and referred to himself as illiterate. In the analysis, this result has been considered insignificant as the majority of interviewees did receive some level of education.

children in a career as a merchant, supplier or skilled worker, with the educational aim that their children have the possibility to have secure job, preferably in a white-collar position, or at least to be employed. The remaining 6.6 percent made no specific comment on their specific desires for their children's jobs or future.

What becomes clear through the surveys is that none of the interviewees envisioned a future in which their children were farmers. Furthermore, there was no apparent correlation between the children's education level and the farmer's own perception on environmental issues. A relationship between the educational background of the children and the environmental awareness and behaviour of the parents could not be made. Thus despite what the more popular theories argue, the survey results showed that among the farmers surveyed, 50 percent wishing for their children to enter into government services, 16.6 percent see no reason for environmental awareness let alone pro-environmental behaviour. In contrast, within the group of farmers hoping to see their children becoming employees or skilled workers (43.3 percent) the level of non-involvement on environmental issues was higher at 20 percent.

These figures suggest that working for the government implies a certain level of environmental understanding, particularly against the backdrop that no matter which office is held, the civil servant encounters environmental laws and regulations, and therefore has to establish environmental awareness. In the case of other job opportunities the perspective of who holds responsibility changes because decisions related to environmental concerns depend on governmental guidelines and are generally made by the management of a company. Thus the employee does not assume liability for any actions at their place of employment that impact that environment.

It is speculation, but significant nonetheless that external factors affect or can bring about individual behaviour and actions (Blake and Fietkau and Kessel), and thus it could be argued that greater responsibility for environmental pollution may develop into environmental awareness. Therefore, this could explain the fact why it would appear that an employee or even employer do not necessarily develop environmental awareness or pro-environmental behaviour. Whether a person develops pro-environmental behaviour or not is to a great extent attributable to individual consequences that the person must face for polluting the environment.[688] Anja Kollmuss and Julian Agyeman have illustrated while utilizing James Blake's theory[689] to argue that a person must overcome three barriers in

---

688 Kollmuss and Agyeman, "Mind the Gap", 239–60.
689 See chapter 1.4.3, Table 5 Blake 1999.

order to transition from environmental concern to pro-environmental behaviour. These three barriers are individuality, responsibility and practicability.[690] It is obvious however that more than half of the farmers in Siziwang are still hindered by these three barriers to act in a pro-environmental manner. This is largely due to laziness and a lack of interest as well as lack of efficacy, which is an effect of the missing property rights and the transfer of responsibility to the local government. It would thus be difficult to hold the farmers in Siziwang solely accountable for the self-made degradation, especially as they see no need for environmental protection. It is not only the practicability factor, which refers to the lack of time, money and information or education on environmental issues and which deters the farmers from pro-environmental behaviour. It is also the first two barriers of individuality and responsibility that seem to be the main reasons for the lack of engagement. This could however change in the near future with the Twelfth Five-Year Plan and its embedded environmental laws and governmental guidelines, which laid the foundation to share accountability with and delegate responsibility to the general population.[691]

8.2.2 Farmers and Herdsmen in Inner Mongolia

The second analysis refers to the evaluation of 254 surveys completed by farmers and herdsmen from different districts in Inner Mongolia. The second set of surveys were held in Inner Mongolia from October to December 2013. Focus was placed on finding a uniform distribution of farmers and herdsmen to build two comparable peer groups for the evaluation of the development of environmental awareness and pro-environmental behaviours in Inner Mongolia. The second narrative did not specifically focus on the agricultural production itself, whereas the Siziwang interviews focused on potato production in order to illustrate concrete challenges facing the farmers and to supply detailed information on environmental problems correlating to farming practices. The goal of the second round of interviews was to gain a wider overview on the general population's environmental perceptions in Inner Mongolia. With the help of students from the Inner Mongolian Agricultural University, the questionnaires were distributed in several districts, in accordance with the students' hometown.

---

690   Ibid.:247
691   Twelfth Five-Year Plan, available at: http://www.britishchamber.cn/content/chinas-twelfth-five-year-plan-2011-2015-full-english-version. Accessed January 2016: 3, 25, 29, 32.

Two groups of 125 farmers and 129 herdsmen were interviewed. Each interviewee provided detailed information on personal background, business knowledge and environmental perception. The evaluation results confirmed that there is no longer an ethnic-economic demarcation line between farmers and herdsmen—Mongolian herdsmen and Han-Chinese farmers—as might have been expected. Today we find a mixed economy and a cultural mosaic of Han-Mongolian families. Within the group of 125 farmers, the distinctive feature shows that 74 percent are of Mongol descent and only 26 percent are Han-Chinese farmers. A little higher is the distribution, which can be found in the evaluation with Inner Mongolian herdsmen. Here the amount of Mongols is 82 percent, but still 16 percent among them are of Han-Chinese ethnicity. Thus what was previously suspected in chapter 3.2.1 appears to be confirmed: Mongolians have become fully integrated, at least in terms of occupation choices, into the Chinese society. Because the traditional Mongolian nomadism continues to be marginalized and often considered inferior, the Mongolian people today appear to typically be involved in the domestic economy working in agriculture and/or commerce. This appears to also be irrespective of the Chinese assimilation policy, which would be an interesting and important subject for further evaluation, though it is not the subject of this work.

In the following graph, a general overview is provided that provides an overview of the Inner Mongolian cropping patterns, which differ from the homogeneous potato cultivation of Siziwang.

Fig. 8: Crop Distribution of Interviewees in Inner Mongolia

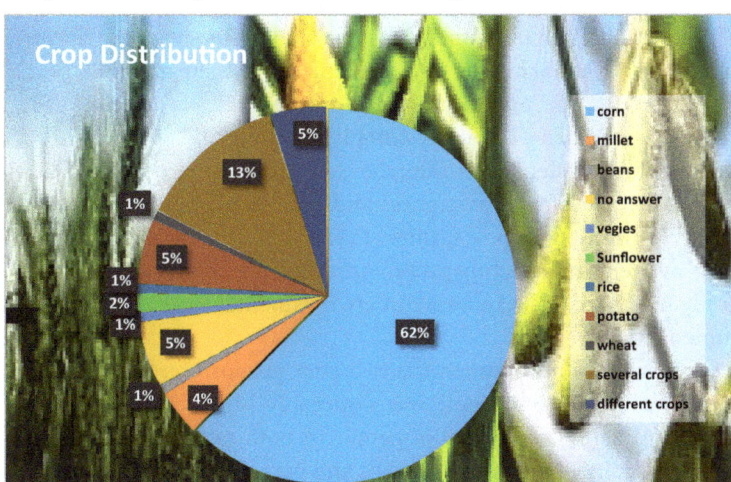

Source: Cilia Neumann (2016)

As can be seen in the diagram, the cultivation of corn is widespread in Inner Mongolia. Slightly more than 60 percent of the farmers are in the corn business and they exclusively cultivate corn. But the cultivation often follows a rotational crop pattern in combination with potato, sunflower, wheat and beans. The farmers, who are mainly cultivating several crops in a rotation, are the second largest group with at a total of 13 percent. Within this group there are farmers cultivating mainly potatoes in combination with other crops like corn, wheat, sunflower and rapeseed. They are mainly situated in the areas of Hohhot, Baotou and Ordos. Some of the farmers polled cultivate millet in combination with corn or broomcorn. Even the alternating cultivation of corn with rice exists, but this pattern is rare and was only found in Tongliao. This combination was also seen in Ulangab but in combination with sunflower cultivation.

Environmental Awareness among Inner Mongolian Farmers and Herdsmen

The correlation between educational achievement and learning about the environment might help to explain the development of environmental awareness that was not that apparent in the surveys completed during this project. Therefore applying the previously mentioned environmental theories of Blake and Fietkau and Kessel is a challenge. Whereas the analysis of the Siziwang potato farmers' interviews clearly show that a higher level of education influences the establishment of a concern for the environment and enables reflection on environmental issues with the aim to act pro-environmentally, the results of the second narrative are very diverse and more difficult to analyze.

The proportional occupational distribution in the survey of farmers and herdsmen is relatively balanced: 49.2 percent are farmers and 50.8 percent are herders (Table 19). The majority in both groups – 58 percent of farmers and 47 percent of herdsmen – typically completed junior or senior middle school. This amount is equivalent with the reference group of farmers and herdsmen disposing of environmental awareness and pro-environmental behaviour (see Table 20). Within the group of farmers (125), 67.2 percent demonstrated environmental awareness and acting pro-environmentally. In the group of herdsmen (129), only 49.6 percent were aware of environmental issues and behaved pro-environmentally.

A dissimilarity found between the two peer groups was that the amount of uneducated herders (32 percent) was noticeably higher than among farmers (20 percent). Whereas the group of people in both peer groups who attended college or university is around 20 percent and here the disparity is marginal.

Tab. 19: Level of Graduation

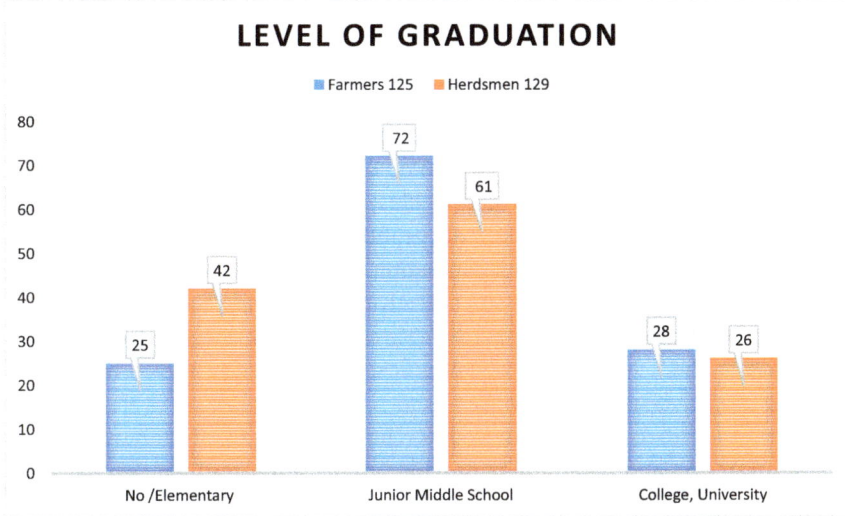

Source: Cilia Neumann 2016, (based on field research Inner Mongolia September 2013) *"Level of Graduation"*

One reason for the educational disparity between uneducated farmers and herdsmen might be that the majority of those questioned were born during the Cultural Revolution when schools were temporarily closed[692] and students were sent to the countryside to work in agrarian brigades to be "re-educated" by peasants about communist ideologies[693]. Today, limited access to educational facilities remains an issue for herders who prefer to live in the more remote parts of

---

692 During the Cultural Revolution, China's educational system came temporarily to a halt. Whereas primary and middle schools were closed at the beginning of the movement and later re-opened, colleges and universities remained closed until 1972. The university entrance exams were only first restored by Deng Xiaoping in 1977.
Joel Andreas, *Rise of the Red Engineers: The Cultural Revolution and the Origins of China's New Class*, Contemporary Issues in Asia and the Pacific, Stanford, CA: Stanford University Press, 2009, 164. http://site.ebrary.com/lib/academiccompletetitles/home.action.

693 At the beginning of *the Down to the Countryside Movement* (上山下乡运动) most of the urban youth volunteered to work in the countryside and "serve" the revolution. Later the people were forced to move to the remote mountainous regions of China. Famous authors, like Jiang Rong, and Zhang Chengzhi, including the Nobel

Inner Mongolia. Even though most of the parents want their children to live an easier life – more successful in economic terms – they are also still trying to sensitize their children to the life of a Mongolian herdsman. Mongolian herders appeared to have kept their deep-rooted nomadic traditions, which are still today quite important as a cultural relict for their pastoral life and the older generation still try to uphold these traditions. Based on the target group of farmers and herders who possess environmental knowledge and simultaneously act pro-environmentally, the following diagram illustrates the educational background of the interviewees. The majority in both groups had an average educational level of a junior or senior middle school graduation. Farmers and herders who did not provide sufficient information to be indicative of possessing an idea of environmentalism or environmental protection awareness have been excluded from this evaluation, as the following analysis specifically seeks to trace the correlation of Awareness-Education Behaviour.

Of the 125 farmers, eighty-four had developed an idea of environmental protection awareness and additionally provided information on their pro-environmental behaviour. Forty-six had completed a junior middle school career, which indicates a general positive result of nearly 55 percent. This amount correlates with the total amount of farmers, who have a junior or senior middle school education, which represented 57.6 percent of the peer group (125 farmers).

Laureate Liu Xiaobo resettled in the countryside of Inner Mongolia and wrote about their experiences during the movement.

Tab. 20: *Correlation of Education to Awareness and Pro-Environmental Behaviour*

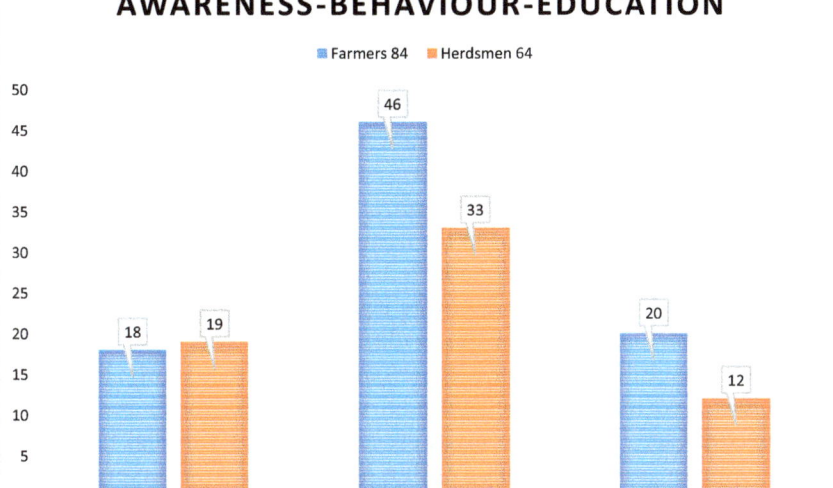

Source: Cilia Neumann 2016, (based on field research Inner Mongolia September 2013), Correlation of Awareness and Pro-Environmental Behaviour (2015)

The number of representatives in the awareness-behaviour reference group of eighty-four farmers with no or only an elementary-level of education is comparatively small (21.4 percent). But the figure is relatively congruent with only 20 percent of farmers with low education levels in the peer group. Nevertheless, in terms of the development of environmental awareness and pro-environmental behaviour, the results are quite surprising: of a total group of twenty-five farmers with low education levels, eighteen had developed environmental awareness and were acting in an environmentally friendly manner, which is an unexpectedly high total of 72 percent within this specific group of farmers. Additional answers regarding environmental understanding may help to clarify the discrepancy between a low education level and a highly sensitive awareness for the environment. All of the remaining eighteen farmers had reflected upon the root causes for Inner Mongolia's environmental crisis though their opinions were deeply divided on this issue. Their beliefs about the causes of the environmental crisis ranged from general climatic conditions (sandstorms and sparse vegetation, temperature rise), overpopulation and incorrect agricultural management (pesticides, fertilizer) to deforestation and industrial development (mining industry and the increase in agricultural land). But when considering alternative prevention methods to stop the ongoing degradation, all farmers agreed and held the same opinion: Only a reduction in farmland while at the same time increasing forests or pastureland through reforestation projects are a sustainable solution. It is difficult to ascertain where in each case this environmental knowledge originates. However, the likely source of this knowledge transfer is the typical exchange that occurs between farmers operating in communities.

The amount of people conscious about environmentalism was surprisingly high in the marginalized groups of well-educated peasants and those with lower levels of education. When looking at the farmers with the highest levels of education, their answers demonstrate more less the same outcomes as the farmers with lower levels of education. Only twenty-eight farmers out of the 125 achieved higher educational qualifications and graduated from colleges or universities. Tellingly, twenty are in the awareness-behaviour group of farmers, who had developed an idea of environmentalism and confirmed the theories of Rajecki, Hines and Hungerford and Tomera: The higher the educational background, the greater the level of the environmental concern. Thus the results of these surveys indicate that environmental awareness and pro-environmental behaviour among farmers is a growing trend. In total, nearly 70 percent are conscious about environmental issues and appear to feel a kind of responsibility for their actions. This is most obvious in the cases where the farming business is more technologically advanced and thus minimizes the environmental impact of agriculture[694]. Still, the disparity between the two target groups, farmers and herdsmen, raises further questions. Whereas nearly three-quarters of farmers appear to have developed ideas about environmentalism, only 50 percent of herders demonstrated a similar awareness of environmental issues. Are these results a reflection of the education levels of the participants or are there other factors involved in this development?

The second target group within the Awareness-Behaviour Education diagram consists of 129 herdsmen of whom sixty-four (50 percent) possessed environmental awareness and pro-environmental behaviour (*Table 20*). Forty-two herders have no or only an elementary education, but within this group nineteen

---

694 The majority of farmers states that reforestation is the key to environmental improvement. Therefore most of the farmers plant trees and support reforestation programs. Environmental protection in private initiative, like saving water, minimizing dirt and using less pesticides is a possibility for some farmers to anticipate their role in pro-environmental behaviour. Sustainable cultivation methods, including responsible use of natural resources, water and land, are seen as particularly promising methods to influence positively the environmental development of Inner Mongolia.

showed an awareness of environmental issues and concerns, and were attempting to act pro-environmentally. This is fewer than within the farmers group, but still 30 percent of the awareness-behaviour reference group. The findings of this evaluation thus seem to support the theoretical approach in which education influences pro-environmental behaviour. Nevertheless, the direct experiences of herdsmen with deteriorated grassland may also have established an ecological sensitivity, which resulted in an increased level of awareness and their pro-environmental behaviour. The diverse answers provided in the interviews[695] trying to explain the formation of environmental problems in Inner Mongolia confirmed that approximately 90 percent of all herdsmen had reflected upon the root causes of grassland degradation and showed awareness for their environment, irrespective of their educational background.[696] But awareness and pro-environmental behaviour had only been developed by 50 percent of herdsmen, which suggests a 40 percent decline in pro-environmental action.

Among the herders with average school career, meaning the completion of junior middle school, (sixty-one people) thirty-three demonstrated environmental awareness and had simultaneously developed pro-environmental behaviour, which corresponds to the 30 percent of the interviewed herdsmen. The majority of herders completed junior and senior middle school (47 percent). Within this class, a little more than half appeared to have reflected on their environment and degradation. This result corresponds with the peer group of 129 herdsmen, wherein 50 percent disposed of environmental awareness and behaviour. Nevertheless there is still a minority of 20 percent of herders who have attained the highest levels of education represented in the surveys having graduated from private colleges or universities. Here the distribution of environmental awareness and behaviour seems to be underrepresented in this group, compared to the relatively high amount of farmers with the same attributes—a ratio of 50 percent to 70 percent.

---

695 Among the 129 herders, who have been interviewed (multiple choices were possible), the majority explains that overgrazing, too many livestock, land reclamation for mining and industrial increase in the grasslands are responsible for the degradation. The second frequently mentioned answer is interesting, as it blames over-population and human expansion into the grasslands as the main cause for environmental problems. Land reclamation for agricultural purpose and cultivation of fields are the third commonly used answer. The rest refers to weather conditions and climatic changes as well as having no idea about environmental problems.

696 Of 129 interviewed herdsmen, 116 speculate about the reasons for the diminishing grassland and the increase of eroded pastures.

Developing environmentalism within the less educated class should be of current interest for local governments as well as international NGOs working on site in Inner Mongolia. The eradication of social stratification is a high priority for the Chinese government and therefore a major aim in the guidelines of the Twelfth Five-Year Plan in China. It remains questionable as to whether or not education is the key to pro-environmental behaviour. Environmental awareness seems to be developed through experiencing environmental issues which hamper the herders' businesses. Severe grassland degradation due to erosion and overgrazing threatens the livestock. Climatic conditions, like the *dzud* are circumstances which frighten the herdsmen, as they have a huge impact on their economic success. It is more likely the fear of economic loss that is the key driver for Inner Mongolian herders to develop environmentalism. Aside from that, it is debatable whether a higher education level leads to more increases in environmental awareness than an average or low- level of education. It seems that the educational background of the parents does have a bearing on the future perspectives or career choices of the children.

Summarizing the Results

The results are remarkably homogeneous between the two subsistence groups (Table 21) including farmers (eighty-four) and herders (sixty-four) who had an idea of environmentalism and behave pro-environmentally. The desire that their children would achieve better job opportunities was well represented in both reference groups: 60 percent of the farmers and herdsmen hoped to see their children in well-paying job positions. Here the question arises as to whether the parents' environmental awareness and the knowledge that economic success in the agricultural sector is highly attributable to a sustainable environment and long-term resource management, may cause some anxiety and restraint regarding the farming and livestock business. As a consequence these attitudes may be responsible for the parents' assumption that higher educational achievement leads to better jobs requiring greater academic qualifications, which in turn support the children becoming independent from the farming sector and moving to urban areas. The opinion that China's metropolises have better future perspectives was previously mentioned in chapter 8.2.1 in the discussion about the retirement plans of potato growers in Siziwang. There the majority also stated that they hoped their children would support them financially in their retirement.

Tab. 21: *Effect of Environmental Awareness to Future Perspective*

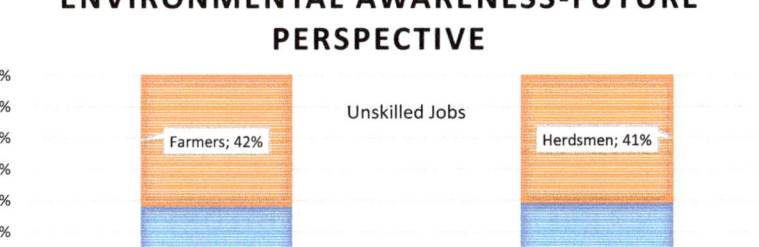

Source: Cilia Neumann 2016, (based on field research Inner Mongolia September 2013) "*Future Perspective*"

Both surveys—the Siziwang potato narrative as well as the questionnaires answered by several Inner Mongolians working in the broad agricultural sector—clearly exemplified that the people in different regions of Inner Mongolia face diverse and severe environmental problems, which are threatening their subsistence lifestyles. The rural exodus of the younger generation is a serious problem for the Inner Mongolian domestic agricultural economy and there are little financial incentives prompting the children of the farmers to carry on their parents' farm or business. Although the majority of the awareness-behaviour reference group was pushing their children into better jobs, the picture changes completely when evaluating the total peer group of 254 participants. The results demonstrate a clear farmer-herder disparity: 56.8 percent of farmers reported looking forward to seeing their children in better jobs, whereas only 47.3 percent of the herdsmen hoped for better future perspectives for their children. Are the parents' life experiences and educational backgrounds influencing their children's school career and choice of occupation?

The following graphic illustrates the parents' level of education as it relates to their future perspective or career choice for their children. Here the underlying values refer to the data of the whole peer group of 254 participants (125/129). There is one significant difference between the two reference groups: If people possessed environmental awareness, which can also be described representatively as a general perception of living conditions, including economic and ecological factors, the more likely they were to be concerned about their children's financial future. Based on this assumption, the previous table illustrated the homogeneity of parental care for the children. The following results, however, would suggest the contrary.

Tab. 22: Parents Career Choices for their Children

CAREER CHOICE FOR CHILDREN

HERDSMEN
- Low Level of Education: 47% | 53%
- Middle Level of Education: 40% | 60%
- High Level of Education: 34% | 66%

FARMER
- Low Level of Education: 52% | 48%
- Middle Level of Education: 65% | 35%
- High Level of Education: 39% | 61%

Source: Cilia Neumann 2016, (based on field research Inner Mongolia September 2013) "*Career Choice*"

Within the group of herders, the parents' whose career choices for their children included qualified jobs were only a small group of the respondents. Remarkably enough, the group which most commonly expressed the desire for their children to attain qualified jobs were the herdsmen with the lowest levels of education and the least educated in the category of the high-level graduates. The general assumption that herdsmen suffer from limited access to education was confirmed when looking at the graduation rates of farmers. Here the amount of well-educated farmers was notably higher, as was the number of their children receiving higher education and better job opportunities, with the notable exception of the farmers with a college or university level of education. Here the majority (61 percent) did not see their children in well-paying jobs, despite their children attending senior middle school or university. Only 39 percent prefer to see their children in advanced job positions. This was an unexpected result that became more understandable through further review of the surveys: The majority of the well-educated farmers with a college- or university-level education were younger than thirty years old and had no children. Thus they did not answer the question on their perspectives for their children's futures. They did however hope for themselves to finish further education and find better job opportunities. Taking this information into account, the illustrated table above supports the thesis of an educational disparity between herders and farmers, which has a great influence on the future perspectives of the next generation. If the urbanization process continues into future the agricultural sector will decrease and monopolies within agriculture will increase. Environmental awareness about grassland degradation or Inner Mongolia's dryland expansion will be replaced by experiences with air and water pollution, emissions and smog in China's most heavily populated areas, where the majority of people tend to live. Nevertheless, the appeal for environmentalism will grow, especially within a new, well-educated generation, whose aim is to support sustainable economy and to reduce environmental damage.

Thus is can be concluded that farmers tended to benefit from higher educational qualifications (80 percent), whereas herders did not (67 percent). This is also in line with levels of environmental awareness and pro-environmental behaviour. It seems then that the theoretical approach in which education plays the key role in developing environmental awareness has been confirmed, although it is more accurate to contribute multiple, rather than a singular factor to the development of environmentalism.

9. Conclusion

It is clear that a variety of causes have led to Inner Mongolia's environmental degradation and how the local populations respond to this degradation is influenced by many different factors. This final chapter is dedicated to returning to the original three questions of this dissertation.

First: How do the past and current food resources meet the demands of China's population? The more general answer is that China has major problems guaranteeing a staple food supply. It is not insignificant that the Chinese government is playing an active role in the current case of "land grabbing" in Africa.[697] This is largely due to China's continuously rising food demand and the aim to remain self-sufficient while at the same time coping with the price competition of globalized markets. These factors are together responsible for pushing the domestic agricultural production. But Inner Mongolia's agricultural production can hardly meet the national demands, especially as it struggles with limited arable land and the lack of sufficient agricultural areas that are suitable for the cultivation of staple crops. Unproductiveness in terms of crop yields, land fragmentation due to smallholders, outdated technologies and poor seed quality complete this critical situation. The combination of the above mentioned restrictions has largely contributed to the current land degradation of Inner Mongolia's agriculture. To cope with the difficult situation, potato cultivation has been proven to be the most successful way for Inner Mongolian drylands to be cultivated effectively. The high-yield potential, water-saving and drought resistant factors of potato cultivation are the main reasons why the potato is being so heavily subsidized by the Chinese government. This also explains why Inner Mongolian farmers have recently come to realize that relying on potatoes can be an opportunity to increase their individual household income. Nevertheless, the Inner Mongolian potato agriculture still faces difficulties to remain competitive in globalized markets. Therefore it is necessary to reform the domestic agriculture. On the one hand it is important that smallholders are incorporated into the domestic economy and that they can gain access to the local markets to sell their products at steady prices. On the other hand, smallholders are incapable of investing on a large scale in modern agricultural technology, as their fields are too small to be cultivated in

---

697 Borras Jr., Saturnino M., and JENNIFER C. FRANCO. "Global Land Grabbing and Trajectories of Agrarian Change: A Preliminary Analysis." *Journal of Agrarian Change* 12, no. 1 (2012): 34–59. doi:10.1111/j.1471-0366.2011.00339.x.

a highly professional or mechanized manner, which would yield enough profit to even be able to re-invest in the high-tech equipment. That is why the average Inner Mongolian potato farmer, as was interviewed in Siziwang, is largely working for self-sufficiency and a potato yield surplus is a "nice to have" agricultural by-product to be sold at local markets, though unfortunately often at very low prices. Beyond this are of course also the various smallholdings in Inner Mongolia that are responsible for extensive land degradation.

An appropriate measure, as mentioned before, to control land fragmentation is farmer consolidation, which is supported by the Chinese government, with the goal to reduce the number of smallholders and increase the acreage in the hands of large companies or farms.[698] But its realization has proven very difficult. Until today, the government has only granted pecuniary reparations for farmers who give up their fields for reforestation and renaturation projects within the Grain for Green policy, which caused a labor problem. The governmental subsidies for reforestation measures provide higher incomes for the farmers than working in agriculture. Thus the few large farms or seed companies in Inner Mongolia lack employees, who are willed to work in the fields.[699] If land consolidation is to be successful and the aim is that large farms cultivate consolidated plots, the smallholders must be given alternative methods to earn an income. In 2003, the WTO initiated, together with the Chinese and Canadian governments, the Small Farmers Adapting to Global Markets (SFAGM) project in China's north-western provinces. This project's greatest success was the cooperation and collective organization that enabled the Inner Mongolian farmers to increase their marketability of the region's potatoes, together described as "good agricultural practices, improved methods, on-farm quality and food safety assurance."[700] International agribusinesses, like Syngenta, have additionally assisted in providing healthy tuber seeds, technological know-how and allocating international cooperation as well as educational training for the farmers. Nevertheless there is a long way to go for the Inner Mongolian agriculture to catch up with the requirements of globalized markets.

Second: How can Inner Mongolia address the problem of its environmental problems which have increased with economic development? Economic pressure accelerates environmental problems, which must be addressed and discussed

---

698 Confidential information from the Interview with the Head of the Agricultural Bureau in Hohhot, see appendix vi.
699 Ibid.:vi.
700 T. Paul Cox. "Small potatoes, big markets in China's Inner Mongolia." Accessed January 4, 2016. Available at: http://www.new-ag.info/en/focus/focusItem.php?a=1057.

in public forums. The emphasis of this dissertation was placed on Inner Mongolia's chronic water shortage and the issue of dryland expansion and desertification as the main concerns related to the on-going soil erosion, salinization and alkalization. Kuznets model that environmental damage first accelerates simultaneously with economic development and later falls when a certain level of quality of life is reached could only be verified for certain polluters like CO^2 emissions.[701] However, large-scale environmental degradation in Inner Mongolia like dryland expansion, shrinking grasslands, soil erosion and water-shortages are obviously still accelerating, despite the per capita income having continuously increased over the last decades and the significant increase in living standards for the majority of people. Nevertheless a healthy environment is only slowly becoming the focus of interest, even though the population has already demanded a clean environment and a reduction of pollution and health risks. This rethinking developed because environmental pollution shifted into the public awareness. People started to perceive degradation and were being affected more and more by environmental problems, which in turn were being discussed publically. Polluted rivers or expanded desertification became an ever-more popular matter of interest in the media. Decisive for this development was, as Kuznet would point out, a steady progression of wealth, which helped this new found sense of environmentalism to arise. As such, based on the results of this study, in can be concluded that the political concept of economic development and poverty alleviation has proven to be successful for Inner Mongolia. The comprehensive model of economic growth as a "pro-poor" incidence positively influenced the economic development of Inner Mongolia. The continuous growth of Inner Mongolia's GDP has also evidently rearranged the composition of the economic pillars: The secondary (production) and tertiary (service) sector have replaced the primary (agriculture) sector, which diminishes in value. Today Inner Mongolia has overcome its status as an undeveloped province and actually ranks fifth within the first ten provinces in terms of per capita GDP[702] Economic growth and poverty alleviation are still priorities for the Chinese government and part of the national policy. But some of the environmental problems, especially the ones affecting the agricultural primary sector in Inner Mongolia, are obviously limiting its further economic development.

---

701   Jalil, Abdul, and Syed F. Mahmud. "Environment Kuznets curve for CO2 emissions: A cointegration analysis for China." *Energy Policy* 37, no. 12 (2009): 5167–72. Accessed November 8, 2015.
702   National Bureau of Statistics of China. Chinas regional economic development: Inner Mongolia per capita GDP in 2013. Available at: http://data.stats.gov.cn/english/swf.htm?m=turnto&id=3. Accessed February 2, 2016.

Therefore the government has included sustainability as one of the top urgencies for China's economic and environmental development in its latest policy guidelines. Additionally, the population has begun to perceive their environment as fragile and is now calling for action to stop on-going degradation.

Third: How have Inner Mongolians developed environmental awareness and contributed to pro-environmental behaviour? Regarding the development of environmental awareness in Inner Mongolia, the current situation is too varied to come to a single conclusion. In general, however, it can be concluded that the addressing of the current environmental problems in Inner Mongolia supported new developments that have enabled ecological sensitivity and sustainability to come into focus within national policies, political education and public thinking. The economic rise of the region was ameliorated due to the remarkable economic development of the country and rural living conditions have improved as a by-product. This has in turn paved the way for the development of environmental awareness. Inner Mongolia's rural infrastructure of townships and villages has also continuously improved. The government has pushed for the expansion of the electricity grid as well as access to primary health care and public services. Even higher education is no longer seen as a privilege of the upper classes and the majority of the people have had the opportunity to obtain some form of education beyond the elementary level. As such, in the course of this dissertation it became clear that a higher educational qualification is definitely advantageous for the emergence of environmental awareness in the theoretical vein of Rajecki, Hines, Hungerford and Tomera, which refer to education as the key to pro-environmental behaviour. Still, the assumption of Fietkau and Kessel as well as Blake that several internal and external factors play a decisive role in establishing pro-environmental behaviour was also confirmed after analyzing the individual circumstances of the interviewees. Especially with regard to the behaviour of Inner Mongolian herdsmen, the individual environmental awareness and the resulting pro-environmental behaviour refers to specific experiences with environmental problems. Perceiving a decline in grassland ecology and the issue of insufficient fodder for the livestock may be the internal factor, governmental incentives for reforestation may act as an external factor. The surveys performed during this dissertation clearly illustrated that the emergence of environmentalism depends on several factors and the momentum of the individual decision-making process. Nonetheless, a clear trend of rising awareness and pro-environmental behaviour to reduce the environmental issues was observable. The government and NGOs are also actively taking part in shaping the public opinion on environmentalism. Public relations activities as well as concrete environmental protection measures supported by transnational ecology groups and non-profit organizations have also helped to bring Inner Mongolia's environmental crisis into the national spotlight.

Outlook

The future perspective for Inner Mongolia's agriculture, both farming and livestock, is uncertain. Currently, rural life presents very few perspectives for the future. Furthermore, brain drain as well as rural exodus are two major inner-political problems facing Inner Mongolia today. The environmental issues in Inner Mongolia are clearly linked to a series of circumstances, including societal factors and underlying driving forces, which have contributed to the current environmental status quo of Inner Mongolia. The empirical research on potato cultivation performed in Inner Mongolia revealed the fears and worries of Inner Mongolian farmers and illustrated some of the environmental endangerments that impede profit-yielding crop production in the region. Looking at the Siziwang field research, the farmers clearly stated that one of the future perspectives of Inner Mongolia's agriculture lies in the domestic potato production, which is financially supported by the Chinese government and is in turn benefitting the farmers. This trend will continue as long as the national policy pushes the domestic potato production as a second staple food to balance serial shortages and world economic food prices. This is from now until 2025, a major part of the Twelfth Five-Year Plan and will likely also be integral to future Five-Year plans.

To clarify, the hypothesis that the nomadic Mongolian lifestyle has had only a minor impact on Inner Mongolia's environmental degradation could not be verified. On the contrary, the specific demographic structure of Han-Mongolian co-existence has shown that each culture has shaped the environment of Inner Mongolia. With the enormous population increase and the national demand for meat, traditional nomadism was forced to catch up with the more industrialized dairy and meat production. Similarly in the farming sector, economic pressure has contributed to certain environmental problems. Overgrazing became a severe issue during the periods of livestock increase in the 1950s and 1960s and during the 1980s and 1990s, when the number of livestock well exceeded the grassland capacity. The case studies revealed that extensive resource extraction in both livestock production and farming has severely increased grassland degradation and desertification in the last decades. Both industrial sectors have tried and are trying to cope with the resulting environmental difficulties. Many Siziwang potato farmers stated that the majority of farmers fully depend on the potato as the most successful crop in terms of its high profitability and efficiency as well as its capability to cope with the hard cultivation conditions in the northern

Chinese drylands. Crop rotation and intercropping with other cash crops is seen as a sustainable method to counteract infertile soils and help the fields to recover at least somewhat between the sowing periods. Dealing with the chronic water-shortage in Siziwang is also a major problem for the farmers and therefore different methods for solving the individual irrigation issues have been discussed and analyzed. Astonishingly, digging new wells is in falling into discredit because this has been shown to make the water problem worse, as it increases the sinking of water tables. Plastic mulch or drip irrigation systems are seen as useful technologies to reduce the problem of water shortage. The individual return on investment (ROI) of the farmers is mainly in new technologies, machines and healthy tuber seeds, all of which improve the potato quality and increase the yield per acre. Although the majority of potato farmers interviewed possess more than double of the average field size of a rural farmer in China, the investment in high-tech machinery is limited due to a lack of funds for investment and sufficient land size necessary to maintain such massive agricultural equipment.

Herders do appear to try to deal with reduced pastures in different ways. In general the surveys showed that the majority of herdsmen possess some environmental protection awareness and perceive the limitations of degraded grasslands. They have had to adjust the amount of animals in accordance with the quality of the grassland, meaning that severely degraded pastures do not carry the same capacity of livestock as unimpaired fertile grasslands. Individual methods to minimize the grazing impact include the reduction of livestock, grassland enclosure and sheep circulation or rotational grazing practices. Still, half of the interviewed herdsmen were convinced that the Inner Mongolian livestock agriculture is responsible for the loss of grassland and the expansion of degraded, eroded and desertificated pastures. These people demonstrated a strong commitment to their environment and are trying to shape and improve the ecological situation by acting pro-environmentally. Even though it is often just a small shed being used to keep the animals separated in order to give the grassland time to recover or they prefer to feed the animals with fodder or storage hay, the Inner Mongolian herdsmen clearly realized that the livestock production not only has to serve national needs, but also relies on a healthy environment.

But the general problem remains the same: Inner Mongolia's agricultural sector is experiencing pressure from all sides. The national economy expects continuous growth and yet the natural resources are exhausted. Despite this, productivity increase remains the top priority for the Chinese government. There is, however, at least a slow change occurring in thinking towards sustainable production methods as well as environmental protection and awareness. However the trend of urbanization is one of the main restrictions of rural development in Inner Mongolia. The aim of the younger generation is to achieve a college or university education and find a well-paying job in one of the urban agglomerations. If the older generation dies or retires to the cities, many smallholders will disappear for lack of a replacement on their farm. On the one hand, this will allow for more room for large-scale farming or dairy production and land consolidation, which at the moment seems to be one of the few measures to slow unproductive agriculture and the spreading land degradation. On the other hand, if farmers and herders leave their homes, fallow fields and grasslands may have the chance to recover or to be incorporated in green belts or protected ecosystems. But as long as the domestic agricultural sector is increasing there is the danger that every hectare will need to be converted into productive fields or rangelands. Today the few environmental shelterbelts, reforestation programs and nature reserves in Inner Mongolia can be attributed to the commitment of non-profit and environmental organizations as well as charity foundations connected to international companies. They are working hard on-site to educate the local population and develop sensitivity towards environmental concerns. But as the program Greening Earth has revealed, several projects stagnate, due to the public's lack of environmental attitude and engagement. Only little progress could be achieved through government-supported renaturation projects, which are already considered an integral part of the comprehensive policy of land consolidation and not necessarily a clear step towards some sort of environmentally-minded thinking. But if this policy plays a strong part in contributing to the protection of large tracts of grassland or dryland and simultaneously is able to establish highly productive agriculture in selected areas, this policy may successfully pioneer a new and sustainable future for Chinese agriculture. Although foreign reports often present the environmental situation in China as being particularly bleak, if the country follows certain steps towards more sustainable practices and development, they will surely be able to successfully attack these problems.

10. Bibliography

Primary Sources

"Baotou Research Institute of Rare Earth." Accessed June 11, 2013.

OECD Review of Agricultural Policies – China. http://dx.doi.org/10.1787/9264012613.

Drylands development and combating desertification: Bibliographic study of experiences in China. FAO environment and energy paper 15. Rome: Food and Agriculture Organization of the United Nations, 1997.

临河市誌 *Linhe shi zhi*. Di 1 ban. [Huhehaote shi]: 内蒙古人民出版社 Nei Menggu ren min chu ban she, 1997.

农业科技通讯. 第1期. 2008.

Agricultural Policy Monitoring and Evaluation 2015. OECD Publishing, 2015. https://books.google.de/books?id=bHtcCgAAQBAJ.

Admin, W. S. "Potatoes and Potato Products Annual_Beijing_China – Peoples Republic of_12-20-2012." Accessed May 12, 2014.

Bamberg, Sebastian, and Guido Möser. "Twenty years after Hines, Hungerford, and Tomera: A new meta-analysis of psycho-social determinants of pro-environmental behaviour." *Journal of Environmental Psychology* 27, no. 1 (2007): 14–25. doi:10.1016/j.jenvp.2006.12.002.

Baosheng Wu, M.ASCE, Guangqian Wang, and and Junqiang Xia. "Case Study: Delayed Sedimentation Response to Inflow and Operations at Sanmenxia Dam." *JOURNAL OF HYDRAULIC ENGINEERING*, 2007, 482–94.

Bettinger, Robert L. *Hunter-gatherers: Archaeological and evolutionary theory*. 2. print. Interdisciplinary contributions to archaeology. New York: Plenum Press, 1992.

Boehm, Daniela. *Agri-environmental decision-making of chinese farmers: Economic, social and cognitive determinants of farmers' nitrogen overuse in Shandong Province*. Stuttgart: ibidem, 2012.

Borras Jr., Saturnino M., and JENNIFER C. FRANCO. "Global Land Grabbing and Trajectories of Agrarian Change: A Preliminary Analysis." *Journal of Agrarian Change* 12, no. 1 (2012): 34–59. doi:10.1111/j.1471-0366.2011.00339.x.

Cease, Arianne J., James J. Elser, Colleen F. Ford, Shuguang Hao, Le Kang, and Jon F. Harrison. "Heavy livestock grazing promotes locust outbreaks by lowering plant nitrogen content." *Science (New York, N.Y.)* 335, no. 6067 (2012): 467–69. doi:10.1126/science.1214433.

Chang Wenhua, Hou Suzhen, Li Xuechun, and Wang Ping. "Analysis of Erosion and Sedimentation in the Inner Mongolian Reach of the Yellow River." *Yellow River* 31, no. 5 (2009): 38–42.

Chaolin, G. U., W. U. Liya, and Ian Cook. "Progress in research on Chinese urbanization." *Frontiers of Architectural Research* 1, no. 2 (2012): 101–49. Accessed June 11, 2013. doi:10.1016/j.foar.2012.02.013.

China Council for International Cooperation on Environment and Development. "Developing Policies for Soil Environmental Protection in China." CCICED Annual General Meeting Unpublished manuscript, last modified December 12, 2013. http://www.cciced.net/encciced/policyresearch/report/201205/P020120529358298439639.pdf.

China Rare Earth Information. " Baiyunebo – The Wealth of China's Rare Earth." Accessed June 11, 2013.

Chung, Shan-Shan, and Monica M.-Y. Leung. "The value-action gap in waste recycling: the case of undergraduates in Hong Kong." *Environmental Management* 40, no. 4 (2007): 603–12. Accessed October 27, 2015. doi:10.1007/s00267-006-0363-y.

Cindy Hurst. "China's Rare Earth Elements Industry: What can the West Learn." Accessed June 11, 2013. http://www.iags.org/rareearth0310hurst.pdf.

–. "The Metal's Edge: The Rare Earth Dilemma: China's Rare Earth Environmental and Safety Nightmare." 2010. Accessed August 10, 2015. http://fmso.leavenworth.army.mil/documents/China's-rare-earth-nightmare.pdf.

"Feeding Nine Billion: The Issues Facing Global Agriculture." Unpublished manuscript, last modified July 22, 2015. https://croplife.org/crop-protection/benefits/.

D. R. Kemp and D. L. Michalk, ed. *Development of sustainable livestock systems on grasslands in north-western China: Proceedings of a workshop held at the combined International Grassland Congress and International Rangeland Conference, Hohhot, Inner Mongolia Autonomous Region, China.* 2008. Accessed March 19, 2014.

Deutsche Bank Research. "Province: Inner Mongolia." Accessed July 22, 2015.

Dinda, Soumyananda. "Environmental Kuznets Curve Hypothesis: A Survey." *Ecological Economics* 49, no. 4 (2004): 431–55. Accessed December 8, 2014. doi:10.1016/j.ecolecon.2004.02.011.

Doczi, Julian, Roger Calow, and Vanessa d'Alancon. "Growing With Less: China's progress in agricultural water management and reallocation." *Development Progress*, 2014; Case study Report Environment. Accessed April 13, 2015.

Fietkau, Hans-Joachim and Hans Kessel, eds. *Umweltlernen: Veränderungsmöglichkeiten des Umweltbewußtseins; Modelle, Erfahrungen.* Schriften des Wissenschaftszentrums Berlin Sozialwissenschaft und Praxis. Internationales Institut für Umwelt und Gesellschaft 18. Königstein/Ts.: Hain, 1981.

Fishbein, Martin, and Icek Ajzen. *Belief, attitude, intention, and behaviour.* Addison-Wesley series in social psychology. Reading, Mass.: Addison-Wesley Pub. Co; Addison-Wesley, 1975.

Foreign Languages Press, ed. *Situation and Policies of China's Rare Earth Industry.* First Edition. Beijing: Foreign Languages Press, 2012. Accessed June 11, 2013. www.china.org.cn.

Frederick, Chris, and Zhang Lei. "China – Peoples Republic of: China To Boost Potato Production and Transform Potato Into Its Fourth Major GrainPotatoes." GAIN Report CH15036 Unpublished manuscript, last modified January 18, 2016. and Potato Products Annual.

Grassi, Sergio. "Chinas Agrarreform – in Zeiten der globalen Finanzkrise." http://library.fes.de/pdf-files/bueros/china/05996.pdf.

HAN, Hong-yun, and Lian-ge ZHAO. "The Impact of Water Pricing Policy on Local Environment – An Analysis of Three Irrigation Districts in China." *Agricultural Sciences in China* 6, no. 12 (2007): 1472–78. doi:10.1016/S1671-2927(08)60010-3.

He, Jie. "Survey on Environmental Kuznets Curve: from the angle of developing countries: Is the Environmental Kuznets Curve hypothesis valid for developing countries? A survey." Unpublished manuscript, last modified January 15, 2015. Workingpaper.

"Inner Mongolia: Market Profile." Unpublished manuscript, last modified July 22, 2015. http://china-trade-research.hktdc.com/business-news/article/Fast-Facts/Inner-Mongolia-Market-Profile/ff/en/1/1X000000/1X07T7RO.htm.

Hou, Xiao-Yan, Feng-Xin Wang, Jiang-Jiang Han, Shao-Zhong Kang, and Shao-Yuan Feng. "Duration of plastic mulch for potato growth under drip irrigation in an arid region of Northwest China." *Agricultural and Forest Meteorology* 150, no. 1 (2010): 115–21. Accessed November 20, 2014. doi:10.1016/j.agrformet.2009.09.007.

Information Office of the State Council. "Situation and Policies of China's Rare Earth Industry." Accessed June 25, 2013.

Ingold, Tim. *The perception of the environment: Essays on livelihood, dwelling and skill.* Repr. London: Routledge, 2008.

Jiang, Hong. *The Ordos Plateau of China: An endangered environment.* UNU studies on critical environmental regions 1035. Tokyo, New York: United Nations University Press, 1999.

Jin, Quan, ed. *Maogaitu Shan xia de nong mu yan ti: Nei Menggu Zhalute Qi Lubei Zhen Baolenggacha diao cha bao gao*. Di 1 ban 46.

Kollmuss, Anja, and Julian Agyeman. "Mind the Gap: Why do people act environmentally and what are the barriers to pro-environmental behaviour?" *Environmental Education Research* 8, no. 3 (2002): 239–60. Accessed April 22, 2015. doi:10.1080/13504620220145401.

Komatsu, Y., A. Tsunekawa, and H. Ju. "Evaluation of agricultural sustainability based on human carrying capacity in drylands – a case study in rural villages in Inner Mongolia, China." *Agriculture, Ecosystems & Environment* 108, no. 1 (2005): 29–43. doi:10.1016/j.agee.2004.12.017.

Lee, R., F. Yu, K. P. Price, J. Ellis, and P. Shi. "Evaluating vegetation phenological patterns in Inner Mongolia using NDVI time-series analysis." *International Journal of Remote Sensing* 23, no. 12 (2002): 2505–12. Accessed June 11, 2013. doi:10.1080/01431160110106087.

Liu, Tuo. "Influence of the Convention to Combat Desertification on forestry in China." In *GLOBAL CONVENTIONS RELATED TO FORESTS*. Edited by A. Perlis. Unasylva 206., 2001. http://www.fao.org/docrep/003/y1237e/y1237e10.htm#P0_0.

Ma, Laurence J. C., Allen G. Noble, and Laurence J. Ma, eds. *The Environment: Chinese and American views (proceedings of the world's first joint U.S.-China symposium; Racine – Wisc., October 13–14, 1978)*. New York: Methuen, 1981.

Mark Giordano, Zhongping Zhu, and Ximing Cai. "Water Management in the Yellow River Basin: Background, Current Critical Issues and Future Research Needs." Accessed June 11, 2013.

Martin Ravallion. "How long will it take to lift one billion people out of poverty?" Unpublished manuscript, last modified December 11, 2014. Policy Research Working Paper.

Meador, Melinda, Zhang Lei, and John Orlowski. "China – Peoples Republic of: Potato Annual 2013." GAIN Report CH13054 Unpublished manuscript, last modified January 18, 2016. http://gain.fas.usda.gov/Recent%20GAIN%20Publications/Potato%20Annual%202013_Beijing_China%20-%20Peoples%20Republic%20of_10-28-2013.pdf.

Meizhen Liu, Gaoming Jiang et al. "The control of land degradation in Inner Mongolia: a case study in Hunshandak Sandland." KSCX1-08-02 Unpublished manuscript, last modified October 14, 2014.

Nallari, Raj, and Breda Griffith. *Understanding growth and poverty // Economic Policies for Growth and Poverty Reduction: Theory, policy, and empirics*. Washington, D.C.: World Bank, 2011.

Nunn, N., and N. Qian. "The Potato's Contribution to Population and Urbanization: Evidence From A Historical Experiment." *The Quarterly Journal of Economics* 126, no. 2 (2011): 593–650. Accessed October 4, 2015. doi:10.1093/qje/qjr009.

OECD. "OECD-FAO Agricultural Outlook 2013–2022: Highlights." 2013. Accessed February 15, 2015.

OECD Publishing. "Environment, Water Resources and Agricultural Policies." Accessed June 11, 2013.

Papademetriou, Minas K., ed. *Workshop to commemorate the International Year of the Potato- 2008: Bangkok, Thailand 6 May 2008*. Food and Agriculture Organization of the United Nations, 2008. Accessed June 4, 2014.

Qu, Dongyu and Kaiyun Xie, eds. *How the Chinese eat potatoes*. Agriculture and food vol. 1. Hackensack, NJ: World Scientific, 2008. http://site.ebrary.com/lib/alltitles/docDetail.action?docID=10688009.

Rajecki, D. W. *Attitudes, themes and advances*. 1st ed. Sunderland, Mass.: Sinauer Associates, 1982.

Rare Earth. "Current situation and outlook of China Rare Earth Industry: 中国稀土信息." *China Rare Earth Information* 18, no. 12 (2012). Accessed June 11, 2013.

Reynolds, J. F., ed. *Global desertification: Do humans cause deserts? [Report of the 88th Dahlem Workshop on Global Desertification: Do Humans Cause Deserts? Berlin, June 10–15, 2001]*. dwr 88. Berlin: Dahlem Univ. Press, 2002.

Ryan Scott, and Zhang Lei. "China – Peoples Republic of: Potatoes and Potato Products Annual." GAIN Report 12076 Unpublished manuscript, last modified January 18, 2016. http://gain.fas.usda.gov/Recent%20GAIN%20Publications/Potatoes%20and%20Potato%20Products%20Annual_Beijing_China%20-%20Peoples%20Republic%20of_12-20-2012.pdf.

Schlütz, Frank. *Palynologische Untersuchungen über die holozäne Vegetations-, Klima- und Siedlungsgeschichte in Hochasien (Nanga Parbat, Karakorum, Nianbaoyeze, Lhasa) und das Pleistozän in China (Qinling-Gebirge, Gaxun Nur): mit 7 Tabellen / Frank Schlütz*. Berlin: Cramer in der Gebr.-Borntraeger-Verl.-Buchh, 1999.

Schrire, Carmel, ed. *Past and present in hunter gatherer studies*. Walnut Creek, Calif: Left Coast Press Inc, 2009. http://site.ebrary.com/lib/alltitles/docDetail.action?docID=10396023.

Simon Kuznets. "Economic Growth and Income Inequality." *The American Economic Review* XLV, no. 1 (March 1955). Accessed January 27, 2016. https://www.aeaweb.org/aer/top20/45.1.1-28.pdf.

Staff, National R. C. *Grasslands and Grassland Sciences in Northern China // Grasslands and grassland sciences in northern China: A report of the Committee on Scholarly Communication with the People's Republic of China, Office of International Affairs, National Research Council.* Washington: National Academies Press; National Academy Press, 1992.

Stephan Klasen. "Economic Growth and Poverty Reduction: Measurement and Policy issues: OECD Development Centre." Unpublished manuscript, last modified December 15, 2014. Working Paper No. 246.

Sun zhong yao. 五原县志 *Wu yuan xian zhi.* 呼和浩特 Hu he huo hao te: 内蒙古人民出版社 Nei meng gu ren min chu ban she, 1996.

Svensson, Jesper. "Development of Water Markets in the Yellow River Basin: A Case-Study of the Ningxia Hui Autonomous Region." 2014. Accessed April 11, 2015.

"Our Crop Focus: Looking at the R&D Innovations behind some of our solutions in Diverse Field Crops and Speciality Crops." Science matters Unpublished manuscript, last modified January 26, 2016. http://www.nxtbook.com/syngenta/Syngenta_Research_and_Development/Autumn_2013/index.php?startid=13#/0. Keeping up to date with Syngenta Research and Development.

Tatsushi Masuda. "Twenty-Year History of an Activity to Prevent Desertification in Inner-Mongolia, China: 内モンゴル沙漠化防止活動の20年." *Journal of Group Dynamics*, 2014, 3–71; 集団力学集団力学. doi:10.11245/jgd.31.3.

Tsunekawa, Atsushi, Guobin Liu, Norikazu Yamanaka, and Sheng Du, eds. *Restoration and Development of the Degraded Loess Plateau, China.* Ecological Research Monographs. Springer Japan, 2014.

U.S. Agency for International Development. "China: From Land Reform to the Single Greatest Poverty-alleviation Achievement in Human History: Briefing Paper." Unpublished manuscript, last modified October 21, 2014. http://pdf.usaid.gov/pdf_docs/PA00J759.pdf.

UN. "Desertification and Drought: Strategic Framework for combating desertification in Germany's development cooperation: UNCCD, MDGs and the Paris Declaration." http://www.un.org/esa/agenda21/natlinfo/countr/germany/desertification.pdf.

"Global Drylands: A UN syste-wide response." Unpublished manuscript. http://www.unccd.int/Lists/SiteDocumentLibrary/Publications/Global_Drylands_Full_Report.pdf.

Vreugdenhil, Dick, John Bradshaw, Christiane Gebhardt, Francine Govers, Mark A. Taylor, Donald K. MacKerron, and Heather A. Ross. *Potato Biology and Biotechnology: Advances and Perspectives*. 1st ed. Burlington: Elsevier Science, 2011.

Wang, Feng-Xin, Shao-Yuan Feng, Xiao-Yan Hou, Shao-Zhong Kang, and Jiang-Jiang Han. "Potato growth with and without plastic mulch in two typical regions of Northern China." *Field Crops Research* 110, no. 2 (2009): 123–29. Accessed November 20, 2014. doi:10.1016/j.fcr.2008.07.014.

Wang, Feng-Xin, Xiu-Xia Wu, Clinton C. Shock, Li-Yun Chu, Xiao-Xiao Gu, and Xuan Xue. "Effects of drip irrigation regimes on potato tuber yield and quality under plastic mulch in arid Northwestern China." *Field Crops Research* 122, no. 1 (2011): 78–84. Accessed November 20, 2014. doi:10.1016/j.fcr.2011.02.009.

Wang, Lunping, Yaxin Chen, and Guofang Zeng. *Nei Menggu he tao guan qu guan gai pai shui yu yan jian hua fang zhi: Irrigation drainage and salinization control in Neimenggu hetao irrigation area*. Di 1 ban. Beijing: Shui li dian li chu ban she, 1993.

Wen Chen. "Economic Growth and the Environment in China: An Empirical Test of the Environmental Kuznets Curve Using Provincial Panel Data." Unpublished manuscript, last modified July 16, 2015.

Williams, Dee Mack. *Beyond great walls: Environment, identity, and development on the Chinese grasslands of Inner Mongolia*. Stanford, Calif: Stanford University Press, 2002.

"Environment, water resources and agricultural policies: Lessons from China and OECD countries; [these proceedings bring together papers from the Workshop on Environment, Resources and Agricultural Policies in China, held in Beijing, on 19–21 June 2006]." Paris.

Xu, He, Jiahong Liu, Dayong Qin, Xuerui Gao, and Jinyue Yan. "Feasibility analysis of solar irrigation system for pastures conservation in a demonstration area in Inner Mongolia." *Applied Energy* 112 (2013): 697–702. doi:10.1016/j.apenergy.2013.01.011.

Xu, Jiongxin. "A Study of Long Term Environmental Effects of River Regulation on the Yellow River of China in Historical Perspective." *Geografiska Annaler. Series A, Physical Geography* Vol. 75, no. 3 (1993): 61–72. http://www.jstor.org/stable/521025.

Xu Kangning, and Wang Jian. "An Empirical Study of A Linkage Between Natural Resource Abundance and Economic Development." *Economic Research Journal* 1 (2006). Accessed July 16, 2015. http://en.cnki.com.cn/Article_en/CJFDTOTAL-JJYJ200601008.htm.

Yi Wang. "Overview of potato production in China: CIP-China Liaison Office in Beijing." Accessed May 12, 2014. http://www.eseap.cipotato.org/MF-ESEAP/Fl-Library/Pto-China.pdf.

Yuan Zhou, Cristianne Close, Marco Ferroni. "Smallholder mapping and the Syngenta Foundation." Accessed January 4, 2016. http://www.syngentafoundation.org/__temp/sfsa_smallholder_mapping.pdf.

Zeng Peiyan. "Strengthen environmental protection and achieving sustainable development." Nairobi, February 21, 2005. Accessed January 27, 2016. http://www.unep.org/gc/gc23/documents/Zeng_Peiyan_speech.pdf.

Zhang, Heling, and Bofu Song. *Seed Potato Production in China*. 300th ed. Huhehaote, Inner Mongolia: Inner Mongolia University Presss, 1992. http://books.google.de/books?id=vSr8k1ypB28C.

Zhang, Lei, and Meiju Wu. "China-Canada "Small Farmers Adapting to Global Markets" Project." AID-FOR-TRADE CASE STORY: CHINA Unpublished manuscript. http://www.oecd.org/aidfortrade/47027097.pdf.

内蒙、巴盟赴河套地区科技兴农联合调研组. "河套农业发展的有效途径." Accessed July 28, 2015. 维普资讯 http://www.cqvip.com.

巴彦淖尔盟志编纂委员会. 巴彦淖尔盟志 *Bayannao'er meng zhi*. Di 1 ban. Huhehaote Shi: 内蒙古人民出版社 Nei Menggu ren min chu ban she, 1997.

张志杰，杨树青，史海滨，马金慧. "内蒙古河套灌区灌溉入渗对地下水的补给规律及补给系数_张志杰." 农业工程学报 第27, 卷第3期 (2011). Accessed November 12, 2014.

杨树青，叶志刚1，史海滨，兰有廷. "内蒙河套灌区咸淡水交替灌溉模拟及预测_杨树青." 农业工程学报 第26, 卷第8期 (2010). Accessed November 12, 2014.

杨根生, 拓万全, 戴丰年, 刘阳宣, 景可, 李炳元 YANG Gensheng, TU O Wanquan, DAI Fengnian, LIU Yangxuan. "风沙对黄河内蒙古河段河道泥沙淤积的影响_杨根生: Contribution of sand sources to the silting of riverbed in Inner Mongolia Section of Huanghe River." 中国沙漠 (*Journal of Desert Research*) Vol. 23, No. 2 (2003): 152–59. Accessed January 29, 2015.

王关区 Wang, Guan Q. "河套草原农业的发展方向和模式: Hetao Agricultural development, direction and investment pattern." 经济社会, 1995, 48–49.

王天顺 Wang, Tianshun. 黄河文明 河套史: *Huang He wen ming He tao shi*. Di 1 ban. Beijing: Ren min chu ban she, 2006.

蔡大旺. "正确认识化学农药的作用及防止滥用化学农药的对策: Cai DW. 2008. Understand the role of chemical pesticides and prevent misuses of pesticides. Bulletin of Agricultural Science and Technology, 1: 36–38." 农业科技通讯, 第1期 (2008年).

Secondary Sources

Sustainable livelihood for rural households: Contributions from rootcrop agriculture UPWARD annual review and planning workshop 19–22 October 1997, Hanoi, Vietnam. Makati City: Users' Perspectives With Agricultural Research and Development, 1998.

黏土礦物之噴施對蔬菜生長與土壤性質的影響. 土壤環境科學系所, 2013.

A. Perlis, ed. *GLOBAL CONVENTIONS RELATED TO FORESTS.* With the assistance of FAO – Food and Agriculture Organization of the United Nations. Unasylva 206. 2001. Accessed February 25, 2016. http://www.fao.org/docrep/003/y1237e/y1237e00.htm.

Amelung, Iwo. *Der gelbe Fluß in Shandong (1851–1911): Überschwemmungskatastrophen und ihre Bewältigung im China der späten Qing-Zeit 7.* Wiesbaden: Harrassowitz, 2000.

Andreas, Joel. *Rise of the red engineers: The Cultural Revolution and the origins of China's new class.* Contemporary issues in Asia and the Pacific. Stanford, Calif: Stanford University Press, 2009.

Ayer de O. Santos, Leinad, ed. *Hydroelectric Dams on Brazil's Xingu River and Indigenous Peoples.* Cambridge, Massachusetts: Cultural Survival, Inc, 1992.

Bawden, Charles R., and Sanders, Alan J. K. *The modern history of Mongolia.* The Kegan Paul library of central Asia. London, New York: Kegan Paul; Columbia University Press [distributor], 2002.

Blackbourn, David, and Udo Rennert. *Die Eroberung der Natur: Eine Geschichte der deutschen Landschaft.* 2nd ed. München: Pantheon, 2008.

Blaikie, P. and H. Brookfield, eds. *Land Degradation and Society.* Taylor & Francis, 2015. https://books.google.de/books?id=P5lGCgAAQBAJ.

Bulag, Uradyn E. *The Mongols at China's edge: History and the politics of national unity.* Lanham, Md: Rowman & Littlefield, 2002.

Chawla, Mukesh, Gordon Betcherman, and Arup Banerji, eds. *From Red to Gray: The "Third Transition" of Aging Populations in Eastern Europe and the Former Soviet Union.* Washington, DC: World Bank, 2012.

Cheng, Zhiming, Mark Wang, and Junhua Chen, eds. *Urban China in the New Era: Market Reforms, Current State, and the Road Forward.* Berlin, Heidelberg, s.l.: Springer Berlin Heidelberg, 2014. http://dx.doi.org/10.1007/978-3-642-54227-5.

Ch'ü, Ko-p'ing, and Woyen Lee. *Managing the environment in China.* 1st ed. Dublin, Ireland: Tycooly International Pub, 1984.

Crossley, Pamela K. *The Manchus.* Cambridge, Mass: Blackwell Publishers, 1997.

Crossley, Pamela K., Helen F. Siu, and Donald S. Sutton, eds. *Empire at the margins: Culture, ethnicity, and frontier in early modern China* 28. Berkeley: University of California Press, 2006.

Dabringhaus, Sabine. *Das Qing-Imperium als Vision und Wirklichkeit: Tibet in Laufbahn und Schriften des Song Yun (1752–1835)* Bd. 69. Stuttgart: Franz Steiner Verlag, 1994.

–. *Territorialer Nationalismus in China: Historisch-geographisches Denken 1900–1949* Bd. 2. Köln: Böhlau, 2006.

Dai, Qing, Patricia Adams, and John Thibodeau. *Yangtze! Yangtze!* English ed. London, Toronto: Earthscan, 1994.

Delang, Claudio O., and Zhen Yuan. *China's Grain for Green Program: A Review of the Largest Ecological Restoration and Rural Development Program in the World*. Cham, s.l.: Springer International Publishing, 2015. doi:10.1007/978-3-319-11505-4. http://dx.doi.org/10.1007/978-3-319-11505-4.

Di Cosmo, Nicola. *Ancient China and its enemies: The rise of nomadic power in East Asian history*. Cambridge, UK, New York: Cambridge University Press, 2002.

Dikötter, Frank. *Mao's great famine: The history of China's most devastating catastrophe, 1958–1962*. 1st U.S. ed. New York: Walker & Co., 2010.

Economy, Elizabeth C. *The river runs black: The environmental challenge to China's future*. Ithaca, NY: Cornell University Press, 2005.

Edmonds, Richard L., ed. *Managing the Chinese environment*. Oxford: Oxford University Press, 2000.

Ellis, J. E. et al. "Dimensions of desertification in the drylands of northern China." In *Global desertification: Do humans cause deserts? [Report of the 88th Dahlem Workshop on Global Desertification: Do Humans Cause Deserts? Berlin, June 10–15, 2001]*. Edited by J. F. Reynolds, 167–90. dwr 88. Berlin: Dahlem Univ. Press, 2002.

Elverskog, Johan. *Our great Qing: The Mongols, Buddhism and the state in late imperial China*. Paperback ed. Honolulu, Hawaii: Univ. of Hawai'i Press, 2008.

Elvin, Mark. *The retreat of the elephants: An environmental history of China*. New Haven: Yale University Press, 2006.

Elvin, Mark, and Ts'ui-jung Liu. *Sediments of time: Environment and society in Chinese history*. Cambridge: Cambridge University Press, 1998.

Fairbank, John K., ed. *The Cambridge History of China*. Cambridge: Cambridge University Press, 1978.

Fernández-Northcote, E. N., Lynn Erselius, and Charlotte Lizárraga. *Proceedings of the international workshop Complementing Resistance to Late Blight*

(Phytophthera infestans) in the Andes, February 13–16, 2001, Cochabamba, Bolivia. Lima, Perú: GILB, International Potato Center, 2003.

Food and Agriculture Organization of the United Nations. *Drylands Development and Combating Desertification: Bibliographic Study of Experiences in China.* Food and Agriculture Organization of the United Nations, 1997. https://books.google.de/books?id=mvdJKQ-GsNYC.

Gernet, Jacques. *Die chinesische Welt: Die Geschichte Chinas von den Anfängen bis zur Jetztzeit.* 1. Aufl. Suhrkamp-Taschenbuch 1505. Frankfurt am Main: Suhrkamp, 1988.

Gitomer, Charles S. *Potato and sweetpotato in China: Systems, constraints, and potential.* Lima, Peru: International Potato Center Regional Office [u.a.], 1996. https://books.google.de/books?id=Zqz7gnJ7V3cC.

Glaeser, Bernhard. *Umweltpolitik in China: Modernisierung und Umwelt in Industrie, Landwirtschaft und Energieerzeugung.* Sozialwissenschaftliche Studien, Bd. 20. Bochum: Brockmeyer, 1983.

–. *Learning from china? Development and environment in third world countries.* [S.l.]: Routledge, 2013.

Goudie, Andrew. *The human impact on the natural environment: Past, present and future.* Seventh ed. Chichester, UK: Wiley-Blackwell, 2013.

Griffin, Keith B. and Zhao Renwei, eds. *The distribution of income in China.* Basingstoke, Hampshire: Macmillan, 1993.

Grumbine, R. E. *Where the dragon meets the Angry River: Nature and power in the People's Republic of China.* Washington, DC: Island Press, 2010.

Gutsch, Alexander, and Gutsch-Heidenreich. *Innovation Wasser: Beispielhafte Projekte aus den Bereichen Gewässer und Wasser 23.* Berlin: Schmidt, 2001.

Herzog, Rolf. *Sesshaftwerden von Nomaden: Geschichte, gegenwärtiger Stand e. wirtschaftl. wie sozialen Prozesses u. Möglichkeiten d. sinnvollen techn. Unterstützung / Rolf Herzog.* Köln: Westdt. Verl, 1963.

Humphrey, Caroline, and David Sneath. *The end of Nomadism? Society, state and the environment in inner Asia.* Knapwell: White Horse Press, 1999.

Julia Bader. "China, autocratic cooperation and autocratic survival: Innauguraldissertation." Accessed June 11, 2013.

Jun, Jjing. *The Temple of Memories: History, Power, and Morality in a Chinese Village.* Stanford, California: Stanford University Press, 1996.

Kagan, K. *The Imperial Moment.* Harvard University Press, 2010. http://books.google.de/books?id=Ai1_5IHQ9vsC.

Khazanov, A. M., and A. Wink. *Nomads in the sedentary world*. RoutledgeCurzon--IIAS Asian studies series. London: Routledge, 2001. https://books.google.de/books?id=-v_RORENFbMC.

Knapp, Karl W., ed. *Environmental Policies and Development. Planning in Contemporary China and Other Essays*. 1974.

Koppel, Bruce, Norton S. Ginsburg, and T. G. McGee, eds. *The Extended metropolis: Settlement transition in Asia*. Honolulu: University of Hawaii Press, 1991.

Lary, D. *The Chinese State at the Borders*. UBC Press, 2011. http://books.google.de/books?id=pFzhWyoYG_wC.

Lattimore, Owen. *The desert road to Turkestan*. Kodansha globe. New York, N.Y: Kodansha International, 1995.

Li, Shi, Hiroshi Satō, and Terry Sicular. *Rising Inequality in China: Challenges to a Harmonious Society*. Cambridge: Cambridge University Press, 2013. http://site.ebrary.com/lib/alltitles/docDetail.action?docID=10729907.

Lin, G.C.S. *Red Capitalism in South China: Growth and Development of the Pearl River Delta*. Vancouver: UBC Press, 2011. https://books.google.de/books?id=ilqVX4c_0kwC.

Lin, Zhibin. *Voluntary resettlement in China: Policy and outcomes of government-organised poverty reduction projects*. Wageningen [Netherlands]: s.n., 2003.

Liu, Dongsheng, and Tung-sheng Liu. *Loess in China*. 2nd ed 5. Beijing, Berlin: China Ocean Press; Springer-Verlag, 1988.

Long, Norman, Jingzhong Ye, and Yihuan Wang, eds. *Rural transformations and development: China in context the everyday lives of policies and people*. Cheltenham, UK, Northampton, MA: Edward Elgar, 2010.

Longworth, John W., ed. *China's rural development miracle: With international comparisons papers presented at an international symposium held at Beijing, China, 25-29 October 1987*. St. Lucia, Qld, Portland, Or: University of Queensland Press; Distributed by International Specialized Bk. Services, 1989.

Longworth, John W., and Gregory J. Williamson. *China's pastoral region: Sheep and wool, minority nationalities, rangeland degradation and sustainable development*. Wallingford: CAB Internat. [u.a.]; CAB Internat, 1993.

Luisito Bertinelli, Eric Strobl and Benteng Zou. "Sustainable Economic Development and the Environment: Theory and Evidence: Core Discussion Paper." 2006. Accessed December 8, 2014.

MacCully, Patrick. *Silenced rivers: The ecology and politics of large dams*. Enl. and updated ed. London: Zed Books, 2001.

Maddison, Angus. *Contours of the world economy, 1–2030 AD: Essays in macroeconomic history*. Oxford, New York: University Press, 2007. http://search.

ebscohost.com/login.aspx?direct=true&scope=site&db=nlebk&db=nlabk&AN=209684.

Marks, Robert B. *Tigers, Rice, Silk, and Silt: Environment and Economy in Late Imperial South China*. Cambridge: Cambridge University Press, 2003.

Milne, John. *Pastoral systems in marginal environments: Proceedings of a satellite workshop of the XXth International Grassland Congress, July 2005, Glasgow, Scotland*. Wageningen [Netherlands]: Wageningen Academic Publishers, 2005.

Mohanty, M., G. Mathew, R. Baum, and R. Ma. *Grass-Roots Democracy in India and China: The Right To Participate*. SAGE Publications, 2007. https://books.google.de/books?id=07qGAwAAQBAJ.

Molle, F. and P. Wester, eds. *River basin trajectories: societies, environments and development*. Wallingford: CABI, 2009.

Peng, Mike W. *Global strategy*. 3rd ed. Mason, Ohio: South-Western, 2014.

Perkins, Dwight H. *Agricultural development in China, 1368–1968*. New Brunswick: AldineTransaction, 2013.

Potter, Sulamith H., and Jack M. Potter. *China's peasants: The anthropology of a revolution*. Cambridge [England], New York: Cambridge University Press, 1990.

Qing, Dai, John Thibodeau, and Philip B. Williams. *The river dragon has come! The Three Gorges dam and the fate of China's Yangtze River and its people*. An East Gate book. Armonk, N.Y.: M.E. Sharpe, ©1998.

Redclift, Michael R. and Ted Benton, eds. *Social theory and the global environment*. Repr. Global environmental change series. London: Routledge, 1997.

Riskin, Carl, Renwei Renwei, and Shi Li, eds. *China's retreat from equality: Income distribution and economic transition*. Asia and the Pacific. Armonk, NY: Sharpe, 2001.

Rummel, R. J. *China's bloody century: Genocide and mass murder since 1900*. New Brunswick, N.J: Transaction Publishers, 2007.

Scholz, Fred, ed. *Nomaden, mobile Tierhaltung: Zur gegenwärtigen Lage von Nomaden und zu den Problemen und Chancen mobiler Tierhaltung; 20 Beiträge*. Berlin: Das Arab. Buch, 1991.

–, ed. *Nomadismus Bibliographie*. Berlin: Das Arabische Buch, 1992.

–. *Nomadismus: Theorie und Wandel einer sozio-ökologischen Kulturweise*. Erdkundliches Wissen 118. Stuttgart: Steiner, 1995.

Scholz, Roland W., ed. *Perspektive Großes Moos: Wege zu einer nachhaltigen Landwirtschaft; Fallstudie '94*. Zürich: vdf Hochschulverl. an der ETH Zürich, 1995.

Shapiro, Judith. *Mao's War against Nature: Politics and the Environment in Revolutionary China*. Studies in Environment and History. Leiden: Cambridge University Press, 2001. Accessed June 11, 2013.

Shi, Xianglin, Vince Castranova, Val Vallyathan, and William G. Perry, eds. *Molecular Mechanisms of Metal Toxicity and Carcinogenesis*. Boston, MA: Springer US, 2001.

Simpson, James R., Xu Cheng, and Akira Miyazaki. *China's livestock and related agriculture: Projections to 2025*. Wallingford, Oxon: CAB International, 1994.

Smil, Vaclav. "Land Degradation in China: An Ancient Problem." In *Land Degradation and Society*. Edited by P. Blaikie and H. Brookfield, 214–22. Taylor & Francis, 2015.

Sneath, David. *Changing Inner Mongolia: Pastoral Mongolian society and the Chinese state*. Reprint. Oxford: Oxford University Press, 2004.

Song, Ligang and Wing T. Woo, eds. *China's dilemma: Economic growth, the environment and climate change*. Canberra, s.l., Washington, D.C., Beijing, China: Social Sciences Academic Press (China); Asia Pacific Press, 2008.

Squires, V. R. *Rangeland Degradation and Recovery in China's Pastoral Lands*. CABI, 2009. https://books.google.de/books?id=2fu6jVgfh-UC.

Stamp, Hans P. *… und weiss wie Alabaster: Eine Kulturgeschichte der Kartoffel*. Neumünster: Wachholtz, 2013.

Stephenson, David, and Margaret S. Petersen. *Water resources development in developing countries* 41. Amsterdam, New York, New York, NY, U.S.A: Elsevier, 1991.

Summers-Smith, J. D. *In search of sparrows*. London: T & A Poyser, 1992.

Taveirne, P. *Han-Mongol Encounters and Missionary Endeavors: A History of Scheut in Ordos (Hetao) 1874–1911*. Leuven University Press, 2004. https://books.google.de/books?id=z2japTNPRNAC.

Thaxton, Ralph. *Catastrophe and contention in rural China: Mao's Great Leap forward famine and the origins of righteous resistance in Da Fo Village*. Cambridge, New York: Cambridge University Press, 2008.

Tilt, Bryan. *The struggle for sustainability in rural China: Environmental values and civil society*. New York: Columbia University Press, 2010.

Turekian, Karl K., ed. *Treatise on Geochemistry: Treatise on Geochemistry*. 2nd edition. Amsterdam: Elsevier Science, 2013.

Veeck, Gregory, Clifton W. Pannell, Christopher J. Smith, and Youqin Huang. *China's geography: Globalization and the dynamics of political, economic, and social change*. 2. ed. Lanham, Md.: Rowman & Littlefield, 2011.

Vengosh, A. "Salinization and Saline Environments." In *Treatise on Geochemistry: Treatise on Geochemistry*. Edited by Karl K. Turekian. 2nd edition, 325–78. Amsterdam: Elsevier Science, 2013.

Wittfogel, Karl A. *Oriental despotism: A comparative study of total power.* 1. ed. Vintage Books V-701. New York NY u.a.: Vintage Books, 1981.

Wu, Harry X., and Christopher Findlay. *China's grain demand and supply: Trade implications.* Chinese Economy Research Unit working papers 97/4. Adelaide: Univ, 1997.

Wu, Harry X. *Reform in China's agriculture: Trade implications.* EAAU briefing paper series Department of Foreign Affairs and Trade, Australia 9. Barton, ACT, 1997.

Yan, Rui-zhen. "Changes in the system of ownership in rural China." In *China's rural development miracle: With international comparisons papers presented at an international symposium held at Beijing, China, 25–29 October 1987.* Edited by John W. Longworth. St. Lucia, Qld, Portland, Or: University of Queensland Press; Distributed by International Specialized Bk. Services, 1989.

Yang, Jisheng, and Hans P. Hoffmann. *Grabstein: Die große chinesische Hungerkatastrophe 1958–1962 = Mùbēi.* Frankfurt am Main: Fischer, 2012. http://www.gbv.de/dms/faz-rez/FD1201210063635021.pdf.

Young, Jason. *China's Hukou System: Markets, Migrants and Institutional Change.* New York: Palgrave Macmillan, 2013. http://gbv.eblib.com/patron/FullRecord.aspx?p=1249600.

Zhongping, Zhu, ed. *Yellow River comprehensive assessment: Basin features and issues collaborative research between International Water Management Institute (IWMI) and Yellow River Conservancy Commission (YRCC).* IWMI working paper 57. Colombo, Sri Lanka: International Water Management Institute, 2003.

Tertiary Sources

Wasserkraft. 1. Aufl. Informationsschriften der VDI-Gesellschaft Energietechnik Teil 3. Düsseldorf: VDI-GET, 1998.

Beer, Bettina, ed. *Methoden und Techniken der Feldforschung.* Ethnologische Paperbacks. Berlin: Reimer, 2003.

China Knowledge Press. *China Business Guide: 2004 Edition.* Singapore: Chinaknowledge Press, 2004. https://books.google.de/books?id=yQGlmWPQFR0C.

Durlauf, Steven N. and Lawrence E. Blume, eds. *The New Palgrave Dictionary of Economics.* Basingstoke: Nature Publishing Group, 2008.

Fan, Shenggen, ed. *The Oxford companion to the economics of China.* 1. ed. Oxford: Oxford University Press, 2014.

Gao, James Z. *Historical dictionary of modern China (1800–1949).* Historical dictionaries of ancient civilizations and historical eras no. 25. Lanham, Md: Scarecrow Press, 2009.

Haarmann, Harald. *Kleines Lexikon der Völker: Von Aborigines bis Zapoteken.* Orig.-Ausg 1593. München: Beck, 2004.

Haller, Dieter and Bernd Rodekohr, eds. *Dtv-Atlas Ethnologie.* Originalausg. München: Deutscher Taschenbuch Verlag, 2005.

Hellbrück, Jürgen, and Elisabeth Kals. *Umweltpsychologie. Basiswissen Psychologie.* Wiesbaden: VS Verlag für Sozialwissenschaften, 2012. Accessed June 18, 2015. doi:10.1007/978-3-531-93246-0.

Hütte, Michael. *Ökologie und Wasserbau: Ökologische Grundlagen von Gewässerverbauung und Wasserkraftnutzung.* 1st ed. Berlin, Wien, Wiesbaden: Parey-Buchverlag [Vieweg], 2003.

Kollmar-Paulenz, Karénina. *Die Mongolen: Von Dschingis Khan bis heute.* Orig.-Ausg 2730 C. H. Beck Wissen. München: C.H. Beck, 2011.

König, Felix v., and Christoph Jehle. *Bau von Wasserkraftanlagen: Praxisbezogene Planungsunterlagen.* 4th ed. Heidelberg: Müller, 2005.

Krieger, M. *Geschichte Asiens: Eine Einführung.* UTB GmbH, 2009. http://books.google.de/books?id=inpzovSWH28C.

Lee, Richard B. and Richard H. Daly, eds. *The Cambridge encyclopedia of hunters and gatherers.* 1. Aufl. reprinted. Cambridge: Cambridge Univ. Press, 2002.

Pálffy, Sándor O. *Wasserkraftanlagen: Klein- und Kleinstkraftwerke* 322. Ehningen bei Böblingen: Expert-Verlag, 1991.

Rössler, Martin. *Wirtschaftsethnologie: Eine Einführung.* 2., überarb. und erw. Aufl. Ethnologische Paperbacks. Berlin: Reimer, 2005.

Rudolph, Jörg, and Thomas Heberer. *China – Politik, Wirtschaft und Gesellschaft: Zwei alternative Sichten.* Sonderausg. für die Zentralen für politische Bildung in Deutschland. Forum hlz. Wiesbaden: Hessische Landeszentrale für Politische Bildung, 2010.

Sanders, A.J.K. *Historical Dictionary of Mongolia.* Scarecrow Press, 2010. http://books.google.de/books?id=5JN83EDDLl4C.

Schmidt-Glintzer, Helwig. *Das alte China: Von den Anfängen bis zum 19. Jahrhundert.* 5. Aufl., Orig.-Ausg. Beck'sche Reihe C.-H.-Beck-Wissen 2015. München: Beck, 2008.

–. *Das neue China: Von den Opiumkriegen bis heute.* 5., überarb. Aufl., Originalausg. Beck'sche Reihe Wissen 2126. München: Beck, 2009.

Sutton, Mark Q., and E. N. Anderson. *Introduction to cultural ecology.* 3. ed. Lanham, Md.: AltaMira, 2014.

Weiers, Michael. *Geschichte der Mongolen* Bd. 586 [i.e. 603]. Stuttgart: W. Kohlhammer, 2004.

Serials

The Economist. "Little Hu and the Mining of the Grasslands: Soaring demand for a region's minerals stirs unrest and brings challenges for a rising political star." July 14th 2014. http://www.economist.com/node/21558605/print.

Andrew Jacobs. "Winter Leaves Mongolians a Harvest of Carcasses." *New York Times, Asia Pacific*, May 19, 2010. South Hangay Province Journal. http://www.nytimes.com/2010/05/20/world/asia/20mongolia.html?_r=0.

Ash, Robert F., and Richard L. Edmonds. "China's Land Resources, Environment and Agricultural Production." *The China Quarterly* 156 (1998): 836. doi:10.1017/S0305741000051365.

Bao, Maohong. "Environmental History in China." *Environment and History* 10, no. 4 (2004): 475–99. doi:10.3197/0967340042772630.

Barthélemy Courmont. "Towards a "Green Detente" between Japan and China? The Case of Cooperation on Reforestation." *Issues & Studies* 51, no. 3 (2015): 29–62.

Barthold, F.K., M. Wiesmeier, L. Breuer, H.-G. Frede, J. Wu, and F.B. Blank. "Land use and climate control the spatial distribution of soil types in the grasslands of Inner Mongolia." *Journal of Arid Environments* 88 (2013): 194–205. Accessed June 11, 2013. doi:10.1016/j.jaridenv.2012.08.004.

Begzsuren, S., J. E. Ellis, D. S. Ojima, M. B. Coughenour, and T. Chuluun. "Livestock responses to droughts and severe winter weather in the Gobi Three Beauty National Park, Mongolia." *Journal of Arid Environments* 59, no. 4 (2004): 785–96. doi:10.1016/j.jaridenv.2004.02.001.

Blake, James. "Overcoming the 'value-action gap' in environmental policy: Tensions between national policy and local experience." *Local Environment* 4, no. 3 (1999): 257–78. doi:10.1080/13549839908725599.

Blanke, Amelia, Scott Rozelle, Bryan Lohmar, Jinxia Wang, and Jikun Huang. "Water saving technology and saving water in China." *Agricultural Water Management* 87, no. 2 (2007): 139–50. doi:10.1016/j.agwat.2006.06.025.

Bocks, Wolfgang. *Pioniergeist: 100 Jahre Wasserkraft aus Laufenburg.* Zürich, Schweiz: Neidhart & Schön AG, 2008.

Brogaard, Sara, and Zhao Xueyong. "Rural Reforms and Changes in Land Management and Attitudes: A Case Study from Inner Mongolia, China." *AMBIO: A Journal of the Human Environment* 31, no. 3 (2002): 219–25. Accessed October 14, 2014. doi:10.1579/0044-7447-31.3.219.

Brown, Colin, and Chan Kai. "Land reform, household specialization and rural development in China." Agricultural and Natural Resource Economics

Discussion Paper 8 Unpublished manuscript, last modified October 16, 2014. http://ageconsearch.umn.edu/bitstream/123790/2/BrownChen.pdf.

Burgess, J., C. M. Harrison, and P. Filius. "Environmental communication and the cultural politics of environmental citizenship." *Environment and Planning A* 30, no. 8 (1998): 1445–60. doi:10.1068/a301445.

Cai, Ximing. "Water stress, water transfer and social equity in Northern China – implications for policy reforms." *Journal of Environmental Management* 87, no. 1 (2008): 14–25. doi:10.1016/j.jenvman.2006.12.046.

Cai D.W. "Understand the role of chemical pesticides and prevent misuses of pesticides." *Bulletin of Agricultural Science and Technology*, 36–38 (2008): 36–38.

Calow, Roger C., Simon E. Howarth, and Jinxia Wang. "Irrigation Development and Water Rights Reform in China." *International Journal of Water Resources Development* 25, no. 2 (2009): 227–48. doi:10.1080/07900620902868653.

Changliang, Xia, and Song Zhanfeng. "Wind energy in China: Current scenario and future perspectives." *Renewable and Sustainable Energy Reviews* 13, no. 8 (2009): 1966–74. doi:10.1016/j.rser.2009.01.004.

Chen, Haiyan, Beisi Jia, and S.S.Y. Lau. "Sustainable urban form for Chinese compact cities: Challenges of a rapid urbanized economy." *Habitat International* 32, no. 1 (2008): 28–40. Accessed June 11, 2013. doi:10.1016/j.habitatint.2007.06.005.

Chen, Jingsheng, Dawei He, and Shubin Cui. "The response of river water quality and quantity to the development of irrigated agriculture in the last 4 decades in the Yellow River Basin, China." *Water Resources Research* 39, no. 3 (2003). doi:10.1029/2001WR001234. http://onlinelibrary.wiley.com/doi/10.1029/2001WR001234/full.

Chen, Liding, Jun Wang, Bojie Fu, and Yang Qiu. "Land-use change in a small catchment of northern Loess Plateau, China." *Agriculture, Ecosystems & Environment* 86, no. 2 (2001): 163–72. doi:10.1016/S0167-8809(00)00271-1.

Chen, Y., and H. Tang. "Desertification in north China: background, anthropogenic impacts and failures in combating it." *Land Degradation & Development* 16, no. 4 (2005): 367–76. doi:10.1002/ldr.667.

Chen L., Zhu W., Wang W., Zhou X., Li W. "Studies on climate change in China in recent 45 years Scholar." *Acta Meteorologica Sinica* 56 (1998): 257–71. Accessed December 14, 2015. https://scholar.google.de/scholar?q=Studies+on+climate+change+in+China+in+recent+45+years+Acta+Meteorologica+Sinica+56&hl=de&as_sdt=0&as_vis=1&oi=scholart&sa=X&ved=0ahUKEwjPlJvOlNvJAhXDkA8KHbX6DR0QgQMIGjAA.

Chengrui, Mei, and Dregne Harold E. "Review Article:Silt and the future development of China's Yellow River." *The Geographical Journal* 167, no. 1 (2001): 7–22.

Chun, Rosa. "Ethical Values and Environmentalism in China: Comparing Employees from State-Owned and Private Firms." *Journal of Business Ethics* 84, no. 3 (2009): 341–48. Accessed June 11, 2013.

Daniel T. Lichter and Janice A. Costanzo. "How do demographic changes affect labor force participation of women?" *MONTHLY LABOR REVIEW Research Summaries*, 1987, 23–28. Accessed July 21, 2015.

David I. Stern. "The Environmental Kuznets Curve." *Departement of Economics*. Accessed December 8, 2014.

Deng, Xi-Ping, Lun Shan, Heping Zhang, and Neil C. Turner. "Improving Agricultural Water Use Efficiency in Arid and Semiarid Areas of China: "New directions for a diverse planet"." Proceedings of the 4th International Crop Science Congress, 26 Sep-1 Oct 2004, Brisbane. Accessed November 11, 2014.

Dennis P. Sheehy. "Grazing management strategies as a factor influencing ecological stability of Mongolian grasslands." *Nomadic Peoples*, no. 33 (1993): 17–30. Accessed July 1, 2013.

Diamond, Norma, Anita Chan, Richard Madsen, Jonathan Unger, William Hinton, Steven W. Mosher, and Jan Myrdal. "Rural Collectivization and Decollectivization in China – A Review Article." *The Journal of Asian Studies* 44, no. 4 (1985): 785. doi:10.2307/2056449.

Du, Jingyuan, and Woodworth Max D. "Irrigation Society in China's Northern Frontier, 1860's-1920's." Accessed June 12, 2013. http://cross-currents.berkeley.edu.

Eggleston, Karen. "China's demographic change in comparative perspective: Implications for labor markets and sustainable development." *Re-evaluating labor market dynamics a symposium sponsored by the Federal Reserve Bank of Kansas City, Jackson Hole, Wyo., Aug. 21–23, 2014*, 2015, 203–31.

Ellis, J. E. et al. "Sustainability of Inner Mongolian Grasslands: Application of the Savanna Model." *Journal of Range Management* 56, no. 4 (2003): 319–28. Accessed March 30, 2014.

Fang, Q. X., L. Ma, T. R. Green, Q. Yu, T. D. Wang, and L. R. Ahuja. "Water resources and water use efficiency in the North China Plain: Current status and agronomic management options." *Agricultural Water Management* 97, no. 8 (2010): 1102–16. doi:10.1016/j.agwat.2010.01.008.

Feng, Zhao-Zhong, Xiao-Ke Wang, and Zong-Wei Feng. "Soil N and salinity leaching after the autumn irrigation and its impact on groundwater in Hetao

Irrigation District, China." *Agricultural Water Management* 71, no. 2 (2005): 131–43. doi:10.1016/j.agwat.2004.07.001.

Feng-Jun Nie, Si-Hong Jiang, Xin-Xu Su, and Xin-Liang Wang. "Geological features and origin of gold deposits occurring in the Baotou–Bayan Obo district, south-central Inner Mongolia, People's Republic of China." *Ore Geology Reviews* 20 (2002): 139–69. Accessed June 11, 2013.

Finlayson, Brian. "Managing Soil Erosion on the Loess Plateau of China to Control Sediment Transport in the Yellow River-A Geomorphic Perspective." 2002. http://www.tucson.ars.ag.gov/isco/isco12/VolumeI/ManagingSoilErosionontheLoessPlateau.pdf.

Fletcher, Joseph. "The heyday of the Ch'ing order in Mongolia, Sinkiang and Tibet." In *The Cambridge History of China*. Edited by John K. Fairbank, 351–408. Cambridge: Cambridge University Press, 1978.

Fu, Bo-jie, Xu-liang Zhuang, Gui-bin Jiang, Jian-bo Shi, and and Y.-h. Lu. "FEATURE: Environmental Problems and Challenges in China." *Environmental Science & Technology* 41, no. 22 (2007): 7597–7602. doi:10.1021/es0726431.

Gang Chen. "The seeds of China's future." *The World Today* February & March 2015 (2015): 36–39. Accessed July 21, 2015.

Gong, Shiyang. "The Role of Reservoirs and Silt-trap Dams in Reducing Sediment Delivery into the Yellow River." *Geografiska Annaler. Series A, Physical Geography* 69, no. 1 (1987): 173–79. http://www.jstor.org/stable/521375.

Grumbine, R. E. "Where the Dragon meets the Angry River: Conservation on the Edge in Yunnan." *Concentric: Library and Cultural Studies* 34, no. 1 (2008): 171–81.

Guo, Xiao j., Yoshihisa Fujino, Satoshi Kaneko, Kegong Wu, Yajuan Xia, and Takesumi Yoshimura. "Arsenic contamination of groundwater and prevalence of arsenical dermatosis in the Hetao plain area, Inner Mongolia, China." In *Molecular Mechanisms of Metal Toxicity and Carcinogenesis*. Edited by Xianglin Shi et al., 137–40. Boston, MA: Springer US, 2001.

Han, G. D., Li, N., Zhao, M. L., Zhang, M., Wang, Z. W., Li, Z. G., Bai, W. J. P., Jones, R., Kemp, D., Takahashi, T., and Michalk, D. "Changing livestock numbers and farm management to improve the livelihood of farmers and rehabilitate grasslands in desert steppe: a case study in Siziwang Banner, Inner Mongolia Autonomous Region." *Development of Sustainable Livestock Systems on Grasslands in north-western China. ACIAR Proceedings*, 2011, 80–96.

He, J., N. J. Kuhn, X. M. Zhang, X. R. Zhang, and H. W. Li. "Effects of 10 years of conservation tillage on soil properties and productivity in the farming-pastoral ecotone of Inner Mongolia, China." *Soil Use and Management* 25, no. 2 (2009): 201–9. doi:10.1111/j.1475-2743.2009.00210.x.

He Jing. "The Consideration of Flood Disasters in Hetao Areas along the Yellow River -An Urgent Matter to Build Daliushu Reservoir." *WATER POWER* 35, no. 4 (2009): 4–7.

Hines, Jody M., Harold R. Hungerford, and Audrey N. Tomera. "Analysis and Synthesis of Research on Responsible Environmental Behaviour: A Meta-Analysis." *The Journal of Environmental Education* 18, no. 2 (1987): 1–8. doi:10.1080/00958964.1987.9943482.

Ho, Peter. "Mao's War against China? The Environmental Impact of the Grain-First Campaign in China." *The China Journal*, no. 50 (2003): 37–60.

Hong Yang, and Xiubin Li. "Cultivated land and food supply in China." *Land Use Policy* 17 (2000): 73–88. Accessed June 11, 2013.

Hooke, Roger L., and José F. Martín-Duque. "Land transformation by humans: A review." *GSA Today* 12, no. 12 (2012): 4–10. doi:10.1130/GSAT151A.1.

Hou, Xiang-Yang, Ying Han, and Frank Y. Li. "The perception and adaptation of herdsmen to climate change and climate variability in the desert steppe region of northern China." *The Rangeland Journal* 34, no. 4 (2012): 349. doi:10.1071/RJ12013.

HOU, Su-zhen, Wen-hua Chang, Ping Wang, Yong TIAN, and Xiao-yan YI. "Fluvial processes in Inner Mongolia reach of the Yellow River." *JOURNAL OF SEDIMENT RESEARCH*, no. 3 (2010): 44–50.

Hu, Wei. "Household land tenure reform in China: Its impact on farming land use and agro-environment." *Land Use Policy* 14, no. 3 (1997): 175–86. doi:10.1016/S0264-8377(97)00010-0.

Hu, Z. M., S. G. Li, J. W. Dong, and J. W. Fan. "Assessment of changes in the state of the rangelands of Inner Mongolia, China between 1998 and 2007 using remotely sensed data." *The Rangeland Journal* 34, no. 1 (2012): 103. doi:10.1071/RJ11055.

Huang, Pingsha, Xiuli Zhang, and Xingdi Deng. "Survey and analysis of public environmental awareness and performance in Ningbo, China: a case study on household electrical and electronic equipment." *Journal of Cleaner Production* 14, no. 18 (2006): 1635–43. doi:10.1016/j.jclepro.2006.02.006.

Huang, Qiuqiong, Scott Rozelle, Jinxia Wang, and Jikun Huang. "Water management institutional reform: A representative look at northern China." *Agricultural Water Management* 96, no. 2 (2009): 215–25. Accessed November 4, 2014. doi:10.1016/j.agwat.2008.08.002.

Hui-mei HAO, and Zhi-yuan REN. "Land Use/Land Cover Change (LUCC) and Eco-Environment Response to LUCC in Farming-Pastoral Zone, China." Accessed June 11, 2013.

Hurst, Cindy. "China's Ace in the Hole: Rare Earth Elements." Accessed June 11, 2013. ndupress.ndu.edu.

J.E. O'Connor and J.E. Costa. "The World's Largest Floods, Past and Present." *Science for a changing world* Circular 1254 (2004). Accessed October 18, 2015.

Jalil, Abdul, and Syed F. Mahmud. "Environment Kuznets curve for CO2 emissions: A cointegration analysis for China." *Energy Policy* 37, no. 12 (2009): 5167–72. Accessed November 8, 2015. doi:10.1016/j.enpol.2009.07.044.

James M. Blaut. "Environmentalism and Eurocentrism." *Geographical Review* 89, no. 3 (1999): 391–408. Accessed June 11, 2013.

Jeremy Swift, and Robin Mearns. "Mongolian pastoralism on the treshold of the twenty-first century." *Nomadic Peoples*, no. 33 (1993): 3–7. Accessed July 1, 2013.

Jia, Guodong, Xinxiao Yu, and Wenping Deng. "Seasonal water use patterns of semi-arid plants in China." *The Forestry Chronicle* 89, no. 02 (2013): 169–77. doi:10.5558/tfc2013-034.

Jiang, Hong. "Grassland management and views of nature in China since 1949: regional policies and local changes in Uxin Ju, inner Mongolia." *Geoforum* 36, no. 5 (2005): 641–53. doi:10.1016/j.geoforum.2004.10.006.

–. "Decentralization, Ecological Construction, and the Environment in Post-Reform China." *World Development* 34, no. 11 (2006): 1907–21. Accessed June 11, 2013. doi:10.1016/j.worlddev.2005.11.022.

Jiang, Yong. "China's water scarcity." *Journal of Environmental Management* 90, no. 11 (2009): 3185–96. Accessed March 13, 2015. doi:10.1016/j.jenvman.2009.04.016. 19539423.

Jing Liu et. al. "Analysis of virtual water flows related to crop transfer and its effects on local water resources in Hetao irrigation district, China, from 1960 to 2008." *Journal of Food, Agriculture & Environment* 11, no. 1 (2013): 682–86. Accessed March 28, 2014.

Johnson, D. L., S. H. Ambrose, T. J. Bassett, M. L. Bowen, D. E. Crummey, J. S. Isaacson, D. N. Johnson, P. Lamb, M. Saul, and A. E. Winter-Nelson. "Meanings of Environmental Terms." *Journal of Environment Quality* 26, no. 3 (1997): 581. doi:10.2134/jeq1997.00472425002600030002x.

Jun Li, Wen, Saleem H. Ali, and Qian Zhang. "Property rights and grassland degradation: A study of the Xilingol Pasture, Inner Mongolia, China." *Journal of Environmental Management* 85, no. 2 (2007): 461–70. doi:10.1016/j.jenvman.2006.10.010.

Kakinuma, Kaoru, Takahiro Ozaki, Seiki Takatsuki, and Jonjin Chuluun. "How Pastoralists in Mongolia Perceive Vegetation Changes Caused by Grazing." *Nomadic Peoples* 12, no. 2 (2008): 67–73. doi:10.3167/np.2008.120205.

Kang, L., X. Han, Z. Zhang, and O. J. Sun. "Grassland ecosystems in China: review of current knowledge and research advancement." *Philosophical Transactions of the Royal Society B: Biological Sciences* 362, no. 1482 (2007): 997–1008. Accessed June 11, 2013. doi:10.1098/rstb.2007.2029.

Kuznets, Simon. "Economics growth and income inequality." *The American Economic Review* 45, no. 1 (1955): 1–28. Accessed December 8, 2014.

Lanza, Guy R. "Where have al the Rivers gone? Silenced Rivers: The ecology and politics of large dams by Patrick McCully." *Bio Science* 45, no. 7 (1997): 460–61.

Leong, CheeKian. "Special economic zones and growth in China and India: an empirical investigation: International Economics and Economic Policy." *Int Econ Econ Policy* 10, no. 4 (2013): 549–67. doi:10.1007/s10368-012-0223-6.

Li, Hongwei, and Xiaoping Yang. "Temperate dryland vegetation changes under a warming climate and strong human intervention — With a particular reference to the district Xilin Gol, Inner Mongolia, China." *Catena* 119 (2014): 9–20. doi:10.1016/j.catena.2014.03.003.

Li, Shi-Zhong, and Catherine Chan-Halbrendt. "Ethanol production in (the) People's Republic of China: Potential and technologies." *Applied Energy* 86 (2009): S162–S169. Accessed January 18, 2016. doi:10.1016/j.apenergy.2009.04.047.

Li, Yue, Jianping Huang, Mingxia Ji, and Jinjiang Ran. "Dryland expansion in northern China from 1948 to 2008." *Advances in Atmospheric Sciences* 32, no. 6 (2015): 870–76. Accessed July 2, 2015. doi:10.1007/s00376-014-4106-3.

Li Tong, Wang Pengtao, Zhao Jingya, and Liang Biao. "Influence of Human Activities to the Environment of the Yellow River Delta and Its Countermeasures." *Yellow River* 32, no. 4 (2010): 70–71.

Lin, George C., and Samuel P. Ho. "China's land resources and land-use change: insights from the 1996 land survey." *Land Use Policy* 20, no. 2 (2003): 87–107. Accessed June 11, 2013. doi:10.1016/S0264-8377(03)00007-3.

Linda Hershkovitz. "Political Ecology and Environmental Management in the Loess Plateau, China." *Human Ecology* 21, no. 4 (1993). Accessed June 11, 2013.

Liu, Can, and Bin Wu. "Grain for Green Programm in China: Policy Making and Implementation?" Briefing Series, China Policy Institute, University of Nottingham, April 2010. https://www.nottingham.ac.uk/cpi/documents/briefings/briefing-60-reforestation.pdf.

Liu, Changming, and Shifang Zhang. "Drying up of the Yellow River: Its impacts and counter-measures." 2001/2002.

Liu, G.Y., Z.F. Yang, B. Chen, Y. Zhang, L.X. Zhang, Y.W. Zhao, and M.M. Jiang. "Emergy-based urban ecosystem health assessment: A case study of Baotou,

Liu, Guo W. "On the geo-basis of river regulation in the lower reaches of the Yellow River." *Science China, Earth Sciences* Vol. 55, no. 4 (2012): 530–44.

Liu, Jianguo, and Jared Diamond. "China's environment in a globalizing world." *Nature* 435, no. 7046 (2005): 1179–86. doi:10.1038/4351179a.

Liu, Jinlong. "Reconstructing the History of Forestry in Northwestern China, 1949–98." *Global Environment 3 (2009): 190–221.* 3 (2009): 190–221. Accessed June 11, 2013.

LIU, Yansui, Jay GAO, and Yanfeng Yang. "A Holistic Approach Towards Assessment of Severity of Land Degradation Along the Great Wall in Northern Shaanxi Province, China." *Environmental Monitoring and Assessment* 82, no. 2: 187–202. doi:10.1023/A:1021882015299.

Liu Changming and Zhang Shifang. "Drying Up of the Yellow River: Its Impacts and Counter-measures: Mitigation and Adaptation Strategies for Global Change 7: 203–214, 2002." *Mitigation and Adaptation Strategies for Global Change* Vol. 7 (2002): 203–14.

LIU Juan-juan, LI Yong-hua, and GAo Duo-feng. "The study on sustainable development after the grain for green project in the agricultural and pasturing interlaced zone-A case of Siziwang Banner in Inner Mongolia." *TERRITORY & NATURAL RESOURCES STUDY*, no. 1 (2009): 39–42.

Lohmar, Bryan, Jinxia Wang, Rozelle Scott, Jikun Huang, and David Dawe. "China's Agricultural Water Policy Reforms: Increasing Investment, Resolving Conflicts, and Revising Incentives." *Agricultural Information Bulletin*, no. 28 (2003): 1–34. Accessed April 1, 2015.

Lonnie K. Stevans and David N. Sessions. "The Relationship Between Poverty and Economic Growth Revisited." *Journal of Income Distribution* 17, no. 1 (2008): 5–20. Accessed July 16, 2015.

Mao, Weining, and Won W. Koo. "Productivity growth, technological progress, and efficiency change in chinese agriculture after rural economic reforms: A DEA approach." *China Economic Review* 8, no. 2 (1997): 157–74. doi:10.1016/S1043-951X(97)90004-3.

Marin, Andrei. "Between Cash Cows and Golden Calves: Adaptations of Mongolian Pastoralism in the 'Age of the Market'." *Nomadic Peoples* 12, no. 2 (2008): 75–101. doi:10.3167/np.2008.120206.

Max Boisot and John Child. "Organizations as Adaptive Systems in Complex Environments: The Case of China." *Organization Science* 10, no. 3 (1999): 237–52. Accessed June 11, 2013.

Mengli Zhao. "Grassland Resource and its Situation in Inner Mongolia, China." *Bull.Facul.Agric.Niigata Univ.*, 58, no. 2 (2006): 129–32; 新潟大学農学部研究報告, 第58 巻 2 号. Accessed February 13, 2014.

Miah, M. D., M. F. Masum, M. Koike, and S. Akther. "A review of the environmental Kuznets curve hypothesis for deforestation policy in Bangladesh." *iForest - Biogeosciences and Forestry* 4, no. 1 (2011): 16–24. doi:10.3832/ifor0558-004.

Michael P. Lawrence. "Damming Rivers, Damning Cultures." *American Indian Law Review* Vol. 30, no. 1 (2005/2006): 247–89. http://www.jstor.org/stable/20070754.

Miller, Daniel, and Dennis Sheehy. "The Relevance of Owen Lattimore's Writings for Nomadic Pastoralism Research and Development in Inner Asia." *Nomadic Peoples* 12, no. 2 (2008): 103–15. doi:10.3167/np.2008.120207.

Mu, S. J., Y. Z. Chen, J. L. Li, W. M. Ju, I. O. A. Odeh, and X. L. Zou. "Grassland dynamics in response to climate change and human activities in Inner Mongolia, China between 1985 and 2009." *The Rangeland Journal* 35, no. 3 (2013): 315. doi:10.1071/RJ12042.

Munasinghe, Mohan. "Is environmental degradation an inevitable consequence of economic growth: Tunneling through the environmental Kuznets curve." *Ecological Economics* 29, no. 1 (1999): 89–109. doi:10.1016/S0921-8009(98)00062-7.

Newhouse, Nancy. "Implications of Attitude and Behaviour Research for Environmental Conservation." *The Journal of Environmental Education* 22, no. 1 (1991): 26–32. Accessed April 22, 2015.

Ou Li, Rong Ma, and James R. Simpson. "Changes in the nomadic pattern and its impact on the Inner Mongolian steppe grasslands ecosystem." *Nomadic Peoples*, no. 33 (1993): 63–72. Accessed July 1, 2013.

Peng, Xiujian. "DEMOGRAPHIC SHIFT, POPULATION AGEING AND ECONOMIC GROWTH IN CHINA: A COMPUTABLE GENERAL EQUILIBRIUM ANALYSIS." *Pacific Economic Review* 13, no. 5 (2008): 680–97. doi:10.1111/j.1468-0106.2008.00428.x.

Peter E. Robertson. "The Global Impact of China's Growth: Economics." In *The Oxford companion to the economics of China*. Edited by Shenggen Fan. 1. ed. Oxford: Oxford University Press, 2014. Accessed February 8, 2016. http://ndc.gov.bd/lib_mgmt/webroot/earticle/611/The-Global-Impact-of-Chinas-Growth.pdf.

Piao, Shilong. "NDVI-indicated decline in desertification in China in the past two decades." *Geophysical Research Letters* 32, no. 6 (2005). doi:10.1029/2004GL021764.

Pietz, D., and M. Giordano. "Managing the Yellow River: continuity and change." In *River basin trajectories: societies, environments and development*. Edited by F. Molle and P. Wester, 99–122. Wallingford: CABI, 2009.

Qian, S., L. Y. Wang, and X. F. Gong. "Climate change and its effects on grassland productivity and carrying capacity of livestock in the main grasslands of China." *The Rangeland Journal* 34, no. 4 (2012): 341. doi:10.1071/RJ11095.

Qiao, Guanghua, Lijuan Zhao, and K. K. Klein. "Water user associations in Inner Mongolia: Factors that influence farmers to join." *Agricultural Water Management* 96, no. 5 (2009): 822–30. doi:10.1016/j.agwat.2008.11.001.

Ravallion, M., and S. Chen. "How Have the World's Poor Fared Since the Early 1980s?" *Policy Research Working Paper 3341, World Bank, Washington, DC*, 2004.

Roberts, Woodford. "Sustainable development, economic growth and environmental degradation: an exploration of their imbrications." Accessed January 15, 2015. https://www.academia.edu/2703507/Sustainable_development_economic_growth_and_environmental_degradation_an_exploration_of_their_imbrication.

Rong, Ma. "Migrant and Ethnic Integration in the Process of Socio-economic Change in Inner Mongolia: a Village Study." *Nomadic Peoples*, no. 33 (1993): 173–91. Accessed January 9, 2015.

Rossabi, M. "Book Reviews: Bagaryn Shirendyb, et al., History of the Mongolian People's Republic. trans. by William A. Brown and Urgunge Onon. Cambridge Mass.: Harvard University Press, 1976. xv + 897 pp. $15.00." *Journal of Asian and African Studies* 15, 1–2 (1980): 176. doi:10.1177/002190968001500126.

Rossabi, Morris. "Bagaryn Shirendyb, et al., History of the Mongolian People's Republic. trans. by William A. Brown and Urgunge Onon. Cambridge Mass: Harvard University Press, 1976. xv + 897 pp. $15.00." *African and Asian Studies* 15, no. 1 (1980): 176. doi:10.1163/156852180X00266.

Roy E. Cordato. "The Polluter Pays Principle: A proper Guide for Environmental Policy." *Institute for Research on the Economics of Taxation Studies in Social Cost, Regulation, and the Environment*, no. 6 (2001): 1–21. Accessed July 16, 2015.

Schneider, K., J.A. Huisman, L. Breuer, and H.-G. Frede. "Ambiguous effects of grazing intensity on surface soil moisture: A geostatistical case study from a steppe environment in Inner Mongolia, PR China." *Journal of Arid Environments* 72, no. 7 (2008): 1305–19. Accessed June 11, 2013. doi:10.1016/j.jaridenv.2008.02.002.

Shang, Ke-Zheng, et al. "Response of climatic change in north China deserted region to the warming of the Earth." *Journal of Desert Research* 21, no. 4 (2001): 387–392.

SHEN, Junyi. "A simultaneous estimation of Environmental Kuznets Curve: Evidence from China." *China Economic Review* 17, no. 4 (2006): 383–94. Accessed July 16, 2015. doi:10.1016/j.chieco.2006.03.002.

SONG, Tao, Tingguo ZHENG, and Lianjun TONG. "An empirical test of the environmental Kuznets curve in China: A panel cointegration approach." *China Economic Review* 19, no. 3 (2008): 381–92. doi:10.1016/j.chieco.2007.10.001.

Soni, Sharad K. "Looking Back to History: Inner Mongolia under Qing Rule." *Bimonthly Journal of Mongolian and Tibetan Current Situation* 蒙藏現況雙月報第十六卷第五期 1, no. 6 (2007): 32–53. Accessed March 18, 2014. http://www.mtac.gov.tw/mtacbook/upload/09609/0701/2.pdf.

Stern, David I. "The Rise and Fall of the Environmental Kuznets Curve." *World Development* 32, no. 8 (2004): 1419–39. Accessed December 8, 2014. doi:10.1016/j.worlddev.2004.03.004.

Stern, David I., Michael S. Common, and Edward B. Barbier. "Economic growth and environmental degradation: The environmental Kuznets curve and sustainable development." *World Development* 24, no. 7 (1996): 1151–60. Accessed December 8, 2014. doi:10.1016/0305-750X(96)00032-0.

SUN, Dan-Feng, Hong LI, R. Dawson, Cheng-Jie Tang, and Xian-Wen Li. "Characteristics of Steep Cultivated Land and the Impact of the Grain-for-Green Policy in China." *Pedosphere* 16, no. 2 (2006): 215–23. Accessed June 11, 2013. doi:10.1016/S1002-0160(06)60046-5.

Susmita Dasgupta, Benoit Laplante, Hua Wang and David Wheeler. "Confronting the Environmental Kuznets Curve." *Journal of Economic Perspectvives* 16, 147–168 (2002). Accessed December 8, 2014.

T. Paul Cox. "Small potatoes, big markets in China's Inner Mongolia." Accessed January 4, 2016. http://www.new-ag.info/en/focus/focusItem.php?a=1057.

Tao, Wang, and Wu Wei. "Combating desertification in China." *Engineering construction* 379 (1998): 50–64.

Tao Yang, Qiang Zhang, and Yongqin David Chen et. al. "A spatial assessment of hydrologic alteration caused by dam construction in the middle and lower Yellow River, China." *HYDROLOGICAL PROCESSES* Vol. 22 (2008): 3829–43.

Tessema, Workneh K., Paul T. M. Ingenbleek, and Hans C. van Trijp. "Pastoralism, sustainability, and marketing. A review: Agronomy for Sustainable Development." *Agronomy for Sustainable Development* 34, no. 1 (2014): 75–92. doi:10.1007/s13593-013-0167-4.

Thomas E. Ewing. "Russia, China, and the Origins of the Mongolian People's Republic, 1911–1921: A Reappraisal." *The Slavonic and East European Review* 58, no. 3 (1980): 399–421. Accessed July 16, 2015.

Tilt, Bryan. "Smallholders and the 'Household Responsibility System': Adapting to Institutional Change in Chinese Agriculture." *Human Ecology* 36, no. 2 (2008): 189–99. doi:10.1007/s10745-007-9127-4.

Tobey, James A., and Henri Smets. "The Polluter-Pays Principle in the Context of Agriculture and the Environment." *The World Economy* 19, no. 1 (1996): 63–87. doi:10.1111/j.1467-9701.1996.tb00664.x.

Topping, Audrey R. "Ecological Roulette: Damming the Yangtze." *Foreign Affaires* Vol. 75, no. 5 (1995): 132–46. http://www.jstor.org/stable/20047305.

van den Bergh, Jeroen C. J. M. "Ecological economics: themes, approaches, and differences with environmental economics." *Regional Environmental Change* 2, no. 1 (2001): 13–23. Accessed January 11, 2015. doi:10.1007/s101130000020.

Varis, Olli, and Pertti Vakkilainen. "China's 8 challenges to water resources management in the first quarter of the 21st Century." *Geomorphology* 41, 2–3 (2001): 93–104. doi:10.1016/S0169-555X(01)00107-6.

Walker, Gordon. "Polluters, victims, citizens, consumers, obstacles, outsiders and experts." *Local Environment* 4, no. 3 (1999): 253–56. Accessed October 27, 2015. doi:10.1080/13549839908725598.

Wang, Guangqian, Baosheng Wu, and Zhao-Yin Wang. "Sedimentation problems and management strategies of Sanmenxia Reservoir, Yellow River, China." *Water Resources Research* 41, no. 9 (2005): n/a. doi:10.1029/2004WR003919.

Wang, Ji-Jun, Zhi-De Jiang, and Zi-Lan Xia. "Grain-for-Green Policy and Its Achievements." In *Restoration and Development of the Degraded Loess Plateau, China*. Edited by Atsushi Tsunekawa et al., 137–47. Ecological Research Monographs. Springer Japan, 2014.

Wang, X. M., C. X. Zhang, E. Hasi, and Z. B. Dong. "Has the Three Norths Forest Shelterbelt Program solved the desertification and dust storm problems in arid and semiarid China?" *Journal of Arid Environments* 74, no. 1 (2010): 13–22. Accessed September 20, 2015. doi:10.1016/j.jaridenv.2009.08.001.

Wang, Xiuhong, Changhe Lu, Jinfu Fang, and Yuancun Shen. "Implications for development of grain-for-green policy based on cropland suitability evaluation in desertification-affected north China." *Land Use Policy* 24, no. 2 (2007): 417–24. Accessed July 2, 2015. doi:10.1016/j.landusepol.2006.05.005.

Wang, Xunming, Fahu Chen, and Zhibao Dong. "The relative role of climatic and human factors in desertification in semiarid China." *Global Environmental Change* 16, no. 1 (2006): 48–57. Accessed July 8, 2015. doi:10.1016/j.gloenvcha.2005.06.006.

Wang, Xunming, Fahu Chen, Eerdun Hasi, and Jinchang Li. "Desertification in China: An assessment." *Earth-Science Reviews* 88, 3-4 (2008): 188-206. doi:10.1016/j.earscirev.2008.02.001.

Wang, Ying, Gustaf Arrhenius, and Yongzhan Zhang. "Drought in the Yellow River – an Environmental Threat to the Coastal Zone." *Journal of Coastal Research*, 2001, 503-15. http://www.jstor.org/stable/25736316.

Webber, Michael, Jon Barnett, Brian Finlayson, and Mark Wang. "Pricing China's irrigation water." *Global Environmental Change* 18, no. 4 (2008): 617-25. doi:10.1016/j.gloenvcha.2008.07.014.

Wei, Liu, Cao Shengkui, Xi Haiyang, and Feng Qi. "Land use history and status of land desertification in the Heihe River basin." *Natural Hazards* 53, no. 2 (2010): 273-90. doi:10.1007/s11069-009-9429-5.

Wen, Zhao. "An Empirical Study of The Linkage Between Resources Development And Economic Development-Taken Shanxi Province as example." *Energy Procedia* 5 (2011): 1394-98. doi:10.1016/j.egypro.2011.03.241.

Wenjun Li, Lynn Huntsinger. "China's Grassland Contract Policy and its Impacts on Herder Ability to Benefit in Inner Mongolia: Tragic Feedbacks." *Ecology and Society* 16, no. 2 (2011): 1-14. Accessed June 8, 2015.

WenJun Zhang, FuBin Jiang, JianFeng Ou. "Global pesticide consumption and pollution: with China as a focus." *International Academy of Ecology and Environmental Sciences* 1, no. 2 (2011): 125-44. Accessed January 11, 2015.

Williams, Dee Mack. "Grassland enclosures." *Human Organization* 55, no. 3 (1996): 307-14. Accessed October 14, 2014.

Wu, Baosheng, Zhaoyin Wang, and Changzhi Li. "Yellow River Basin Management and current issues." *Journal of Geographical Sciences* Vol. 14 (2004): 29-34. http://www.geog.cn.

Wu, Bo, and Long J. Ci. "Landscape change and desertification development in the Mu Us Sandland, Northern China." *Journal of Arid Environments* 50, no. 3 (2002): 429-44. Accessed June 11, 2013. doi:10.1006/jare.2001.0847.

Wu, Ziping, Minquan Liu, and John Davis. "Land consolidation and productivity in Chinese household crop production." *China Economic Review* 16, no. 1 (2005): 28-49. Accessed September 30, 2014. doi:10.1016/j.chieco.2004.06.010.

Xepapadeas, Anastasios. "ecological economics." In *The New Palgrave Dictionary of Economics*. Edited by Steven N. Durlauf and Lawrence E. Blume, 599-604. Basingstoke: Nature Publishing Group, 2008.

Xiao, Xiangming, Stephen Boles, Jiyuan Liu, Dafang Zhuang, Steve Frolking, Changsheng Li, William Salas, and Berrien Moore. "Mapping paddy rice agriculture in southern China using multi-temporal MODIS images." *Remote*

Sensing of Environment 95, no. 4 (2005): 480–92. Accessed July 21, 2015. doi:10.1016/j.rse.2004.12.009.

Xie, Yina, and Wenjun Li. "Why do Herders Insist On Otor? Maintaining Mobility in Inner Mongolia." *Nomadic Peoples* 12, no. 2 (2008): 35–52. doi:10.3167/np.2008.120203.

Xing, Li, and Qi Xing. *Nei Menggu qu yu you mu wen hua de bian qian: The Nomadic Cuture Change in Inner Mongolia.* Di 1 ban. Beijing, 2013.

Xiuxia, Bao, Yi Jin, Liu Shurun, Gaowa Jimuse, Wang Puchang, and Lian Yong. "Effects of Different Grazing on the Typical Steppe Vegetation Characteristics on the Mongolian Plateau." *Nomadic Peoples* 12, no. 2 (2008): 53–66. doi:10.3167/np.2008.120204.

Xu, Beina. *China's Environmental Crisis.* 2014. http://www.cfr.org/china/chinas-environmental-crisis/p12608.

Xu, Jianchu, Yong Yang, Zhuoqing Li, Nyima Tashi, Rita Sharma, and Jing Fang. "Understanding land use, livelihoods, and health transitions among Tibetan nomads: a case from Gangga Township, Dingri County, Tibetan Autonomous Region of China." *EcoHealth* 5, no. 2 (2008): 104–14. doi:10.1007/s10393-008-0173-1.

Xu, Jiongxin. "Naturally and Anthropogenically accelerated sedimentation in the Lower Yellow River, China, over the Past 13.000 years." *Geografiska Annaler. Series A, Physical Geography* Vol. 80 A, no. 1 (1998): 67–78.

–. "Historical bank-breaching of the lower Yellow River as influenced by drainage basin factors." *Catena* Vol. 45 (2001): 1–17. http://www.elsevier.com/locate/catena.

–. "Growth of the Yellow River Delta over the past 800 Years, as Influenced by Human Activities." *Geografiska Annaler. Series A, Physical Geography* Vol. 85, no. 1 (2003): 21–30.

Xu, Xu, Guanhua Huang, Zhongyi Qu, and Luis S. Pereira. "Assessing the groundwater dynamics and impacts of water saving in the Hetao Irrigation District, Yellow River basin." *Agricultural Water Management* 98, no. 2 (2010): 301–13. Accessed June 12, 2013. doi:10.1016/j.agwat.2010.08.025. http://www.elsevier.com/locate/agwat.

Xu Jiongxin. "A study of sediment delivery by floods in the lower Yellow River, China." *Hydrological Sciences Journal* 48, no. 4 (2003): 553–66. doi:10.1623/hysj.48.4.553.51416.

Yang, Hong, Xiaohe Zhang, and Alexander J. Zehnder. "Water scarcity, pricing mechanism and institutional reform in northern China irrigated agriculture." *Agricultural Water Management* 61, no. 2 (2003): 143–61. doi:10.1016/S0378-3774(02)00164-6.

Yang, Jiawen, and Ralph Gakenheimer. "Assessing the transportation consequences of land use transformation in urban China." *Habitat International* 31, 3-4 (2007): 345-53. Accessed June 11, 2013. doi:10.1016/j.habitatint.2007.05.001.

Yang, Lijing, Yonghong Niu, and Yanli Xu. "Sustainable Development and Formation of Harmonious Nature." *Energy Procedia* 5 (2011): 629-32. Accessed June 11, 2013. doi:10.1016/j.egypro.2011.03.110.

Yang, X., Z. Ding, X. Fan, Z. Zhou, and N. Ma. "Processes and mechanisms of desertification in northern China during the last 30 years, with a special reference to the Hunshandake Sandy Land, eastern Inner Mongolia." *Catena* 71, no. 1 (2007): 2-12. doi:10.1016/j.catena.2006.10.002.

Yang, X., K. Zhang, B. Jia, and L. Ci. "Desertification assessment in China: An overview." *Journal of Arid Environments* 63, no. 2 (2005): 517-31. doi:10.1016/j.jaridenv.2005.03.032.

Yang, Yanzhao, Zhiming Feng, He Q. Huang, and Yaoming Lin. "Climate-induced changes in crop water balance during 1960-2001 in Northwest China." *Agriculture, Ecosystems & Environment* 127, 1-2 (2008): 107-18. doi:10.1016/j.agee.2008.03.007.

Yang, Yuting, Songhao Shang, and Lei Jiang. "Remote sensing temporal and spatial patterns of evapotranspiration and the responses to water management in a large irrigation district of North China." *Agricultural and Forest Meteorology* 164 (2012): 112-22. doi:10.1016/j.agrformet.2012.05.011.

Yang, Z. F., T. Sun, B. S. Cui, B. Chen, and G. Q. Chen. "Environmental flow requirements for integrated water resources allocation in the Yellow River Basin, China." *Communications in Nonlinear Science and Numerical Simulation* 14, no. 5 (2009): 2469-81. doi:10.1016/j.cnsns.2007.12.015.

Ye, Qian, and Michael H. Glantz. "The 1998 Yangtze Floods: The Use of Short-Term Forecasts in the Context of Seasonal to Interannual Water Resource Management: Mitigation and Adaptation Strategies for Global Change." *Mitig Adapt Strat Glob Change* 10, no. 1 (2005): 159-82. doi:10.1007/s11027-005-7838-7.

Ye, Yu, and Xiuqi Fang. "Boundary shift of potential suitable agricultural area in farming-grazing transitional zone in Northeastern China under background of climate change during 20th century." *Chinese Geographical Science* 23, no. 6 (2013): 655-65. doi:10.1007/s11769-013-0638-1.

You, Liangzhi, Max Spoor, Ulimwengu, John, and Shemai Zhang. "Land use change and environmental stress of wheat, rice and corn production in China." *China Economic Review* 22 (2011): 462-73.

Yu, Ruihong, Tingxi Liu, Youpeng Xu, Chao Zhu, Qing Zhang, Zhongyi Qu, Xiaomin Liu, and Changyou Li. "Analysis of salinization dynamics by remote sensing in Hetao Irrigation District of North China." *Agricultural Water Management* 97, no. 12 (2010): 1952–60. Accessed June 11, 2013. doi:10.1016/j.agwat.2010.03.009.

YU Ruihong, LI Changyou, LIU Tingxi, and XU Youpeng. "The environment evolution of Wuliangsuhai wetland." *Journal of Geographical Sciences* 14, no. 4 (2004): 456–64. Accessed June 11, 2013.

Zha, Yong, and Jay GAO. "Characteristics of desertification and its rehabilitation in China." *Journal of Arid Environments* 37, no. 3 (1997): 419–32. doi:10.1006/jare.1997.0290.

Zhang, Dian, and Changxing Shi. "Sedimentary Causes and Management of Two Principal Environmental Problems in the Lower Yellow River." *Environmental Management* 28, no. 6 (2001): 749–60.

Zhang, MunkhDalai A., Elles Borjigin, and Huiping Zhang. "Mongolian nomadic culture and ecological culture: On the ecological reconstruction in the agro-pastoral mosaic zone in Northern China." *Ecological Economics* 62, no. 1 (2007): 19–26. doi:10.1016/j.ecolecon.2006.11.005.

Zhang, Y., Y. Luo, W. Zhao, and Hatina M. "Land use dynamics and landscape change pattern in Hetao irrigation district." *The International Society for Optical Engineering* 7841 (2010): 1–9.

Zhang, Dongsheng Ma, Xiongxi Hu, Hui. "Arsenic pollution in groundwater from Hetao Area, China." *Environmental Geology* 41, no. 6 (2002): 638–43. Accessed July 28, 2015. doi:10.1007/s002540100442.

Zhao, H.-L., X.-Y. Zhao, R.-L. Zhou, T.-H. Zhang, and S. Drake. "Desertification processes due to heavy grazing in sandy rangeland, Inner Mongolia." *Journal of Arid Environments* 62, no. 2 (2005): 309–19. doi:10.1016/j.jaridenv.2004.11.009.

Zhao, Hong, You-Cai Xiong, Feng-Min Li, Run-Yuan Wang, Sheng-Cai Qiang, Tao-Feng Yao, and Fei Mo. "Plastic film mulch for half growing-season maximized WUE and yield of potato via moisture-temperature improvement in a semi-arid agroecosystem." *Agricultural Water Management* 104 (2012): 68–78. Accessed November 24, 2014. doi:10.1016/j.agwat.2011.11.016.

Zhao, Ying, Stephan Peth, Rainer Horn, Julia Krümmelbein, Bettina Ketzer, Yingzhi Gao, Jose Doerner, Christian Bernhofer, and Xinhua Peng. "Modeling grazing effects on coupled water and heat fluxes in Inner Mongolia grassland." *Soil and Tillage Research* 109, no. 2 (2010): 75–86. Accessed June 11, 2013. doi:10.1016/j.still.2010.04.005.

Zhizhong, Wu, and Du Wen. "Pastoral Nomad Rights in Inner Mongolia." *Nomadic Peoples* 12, no. 2 (2008): 13–33. doi:10.3167/np.2008.120202.

Zhou, Yang, Ning Li, Wenxiang Wu, Jidong Wu, Xiaotian Gu, and Zhonghui Ji. "Exploring the characteristics of major natural disasters in China and their impacts during the past decades." *Natural Hazards* 69, no. 1 (2013): 829–43. Accessed October 18, 2015. doi:10.1007/s11069-013-0738-3.

Zhu, T. X., and A. X. Zhu. "Assessment of soil erosion and conservation on agricultural sloping lands using plot data in the semi-arid hilly loess region of China." *Journal of Hydrology: Regional Studies* 2 (2014): 69–83. doi:10.1016/j.ejrh.2014.08.006.

Zhu Xiaodong. "Understanding China's Growth: Past, Present and Future: Past, Present, and Future." *The Journal of Economic Perspective* 26, no. 4 (2012): 103–24. doi:10.2469/dig.v43.n2.37.

Appendices

Appendix 1 Quantitative Farmers Questionnaire ii
Appendix 2 Quantitative Livestock Breeders Questionnaire iv
Appendix 3 Interview with the Head of the Agricultural Bureau in Hohhot vi
Appendix 4 Interview with Yun Ting, Potato Breeder in Siziwang Qi viii
Appendix 5 Interview with a Herdsman from Siziwang Qi xii
Appendix 6 Interview with a Taxi Driver form Dengkou, Hetao xvi

Appendix 1

Quantitative Farmers Questionnaire

内蒙古_____地区农民调查问卷：

个人信息：

1. 年龄：_____Age 性别：_____Sex 民族：_____Ethnicity
2. 教育程度：_____Education 语言：_____Languages
3. 您在西子王期地区的居住时间：_____How long have you lived in this area?
4. 您的职业：_____Your occupation?
5. 您的家庭有几口人：_____How many people live in your house hold?
6. 您的子女数量：_____How many children do you have?
   他/她们的受教育程度：_____What is their education?
   您希望他/她们将来从事的工作:_____What do you want them to do?
7. 您是户主吗？_____Are you the head of your household?

土地用途

9. 所有的土地总面(亩):_____Total area of all land (*mu*)
10. 总耕地面积:_____Total area of cultivated land (Chinese unit: *mu*)
11. 您培养什么样的庄稼?_____What kind of crops do you plant?
a) 为什么您培养这个庄稼? Why did you choose this crop/fruit?
b) 您从什么时候开始培育您的土地? Since when do you cultivate your land?
c) 您是否每年都会种植不通的植物? Do you change your fruits/crops? (Or Monoculture)
12. 每年家庭得到的政府补贴是多少？_____What is the annual household total subsidy?
13. 每年家庭农业净收入？_____What is the annual household agricultural net income?
14. 您会再投资多少金钱在您的事业? How much money do you reinvest into your business?
a.) 目标? For what?
15. 您是否恐惧环境退化? Do you fear environmental degradation?

16. 您恐惧什么样的环境退化多数？干旱/盐碱化/水土流失? What kind of degradation you fear most?
17. 您认为什么是环境退化的原因? What do you think are the reasons for environmental degradation?
18. 您在防止环境退化方面的尝试是什么? What do you do to prevent further degradation?

Appendix 2

Quantitative Livestock Breeders Questionnaire

内蒙古_____地区牧民调查问卷：

个人信息：

1. 年龄：_____Age 性别：_____Sex 民族：_____Ethnicity
2. 教育程度：_____Education 语言：_____Languages
3. 您在_____地区的居住时间：_____How long have you lived in this area?
4. 您的职业：_____Your occupation
5. 您的家庭有几口人：_____How many people live in your household?
6. 您的子女数量：_____How many children do you have?
   他/她们的受教育程度：_____What is their education?
   您希望他/她们将来从事的工作:_____What do you want them to do?
7. 您是户主吗？_____Are you the head of your household?

土地用途

9. 所有的土地总面(亩)：_____ Total area of all land (*mu*)
10. 放牧的土地总面积：_____ Total area of grazing land
12. 您饲养什么样的家畜? What kind of livestock do you raise?
    a) 为什么您饲养这个家畜? Why did you raise these animals?
    b) 您从什么时候饲养您的家畜? Since when do you raise animals?
    c) 您是否每年都会养殖不通的家畜? Did you change your livestock? (Goat to sheep or cattle?)
13. 你住在村庄或者你还是游牧人吗? Did you settle down or do you still move with the herds?
    a.) 从您居住的地方到牧场有多远? How far are your rangelands?
    b.) 冬季牧场和夏季牧场距离多远? What is distance between the winter and summer rangelands?
    c.) 在冬天您怎么样喂您的动物? How do you feed your livestock in winter?
    d.) 您是否恐惧冻灾，上一次的冻灾是什么时候? Do you fear dzud? And when was the last dzud?
    e) 多少的动物死了?

环境

14. 您是否恐惧草原退化? Do you fear grassland degradation?
15. 您认为什么是草原退化的原因? What do you think are the reasons for grassland degradation?
16. 您为防止草原退化所做的努力? What do you do to prevent further grassland degradation?
17. 饲养家畜的每年家庭收入? What is the annual household income from livestock?
18. 您会再投资多少在您的事业上? How much money do you reinvest into your business?
    a.) 目标? For what?

Appendix 3

Interview with the Head of the Agricultural Bureau in Hohhot

Participants: Dr. Christoph Neumann, David Liu (Syngenta)

Crop Distribution in Inner Mongolia:

1.2 Million hectares land are cultivated acreage of Inner Mongolia.
Total output 25 Mill. Metric tons of field crops, 10% of the total crop production of China
Key crops: corn, soya, wheat, potato and rice
Corn 55% (45 million of mu9)
3rd important crop is potato (10 million mu)
Soya 2nd important crop 20 Million mu
Cereals 4th important 8 million mu
Potato: 18% of Chinas total potato acreage (total 5 million hectares of potato)

Why focus on the potato cultivation in Inner Mongolia?

Competitive advantage in the north for climate, and seed tuber production
Total acreage 2013 10.2 million mu of potato (reason very dry, therefore potato cultivation, and prices rose)
Average yield 1 metric ton per mu (1 mu – 15 ha)
10 million metric tons per year
Ministry says best conditions in Inner Mongolia for potato production (low pests)
Promotion of irrigation: 3–4 metric tons per mu of yield increase, pivot irrigation forbidden (from above), no new pivot systems are built, but old ones can be still used.
Drip Irrigation forced and promoted, Drills are built, but not on spots where water table already decreased.
Potato one of the priority crops is the agenda, including sheep and others.
A lot of governmental benefits for potato production, seed tuber production supported and the government realizes the seed tuber market and production as a key issue for the future to increase the yield.
Inner Mongolia's seed tuber production has technology problems, pests and disease management has to improve, storage and tuber production.
FARMER CONSOLIDATION: small holders are a problem (less than 1 mu), no efficient production, just for subsistence!

Government encourages land aggregation and modernization. Goal is fewer farmers, but bigger acreage in the hold of large companies, consolidation.

Labor problem, government supports reforestation programs and pays higher salaries, therefore potato producers have not enough employers to work on the fields.

2 policies for consolidation: Higher subsidies for families or farms working together/subsidies for virus-free tubers (certified).

4.6 mu per farmer in average (not 0.5 hectare per pax.) compared to 1850 in Europe, the farmer had a bigger acreage, but for China it is still big.

In the EAST: Promotion of livestock

In the WEST: Promotion of staple foods.

Appendix 4

Interview with Yun Ting, Potato Breeder in Siziwang Qi

With Yun Ting, General Manager of *Zhengfeng Seed Potato*
Location: Inner Mongolia Zhengfeng Seed Potato Co. Ltd, Huhhot, China, No. 22, Zhaojun Road, Yuquan, Huhhot, Inner Mongolia, China
Participants: Yun Ting, Wang Yifu, Neumann Cilia

姓名：云挺
性别：男
出生：1963年10月
居住地：呼和浩特
教育程度：硕士
家庭成员：4人
孩子数量：*2 分别在国外读大学和硕士*
工作：*土豆种植*
从事该工作的原因：*开始就业于农业厅，是工作需要*
从事该工作的时间：*15年*
从事该工作的目标：*把公司发展壮大，进一步发展种薯市场和国际合作*
从事该工作初期所面临的困难：*市场不稳定、不规范*
雇佣的员工：60人
谁购买您的产品：*各地的农民、种植大户和种薯经销商*
耕种的土地来源：*向农民租用承包*
土地合约：*10年-20年不等，15年左右的数量较多*
土地灌溉方式：*指针式喷灌以及滴灌*
农业设备的产权：*公司*
您是否认为您的工作是内蒙古发展的一部分：*是的，土豆对于内蒙古农业很重要*
您是否认为您的工作是中国发展的一部分：*是的，我们向全国推广土豆*
您的父母从事何种工作：*农民，种植业*
他们面临的困难：*年龄大了，不再从事农业生产*
您认为河套的环境如何：*农业种植环境变好，盐碱土地变良田，灌溉地区增加*
环境变化对于您事业的影响：*农业生产永远离不开环境变化，农业生产需要承担双重的风险，市场的风险和自然的风险.*
您对于环境的看法：*需要创造好的生态环境，这是每个人都应该有的意识；要节约用水，合理利用水资源，改进灌溉方式.*
您对于环境改善的努力：*合理利用水资源，节约用水；为了保护土壤，增加生物肥料和菌类肥料.*

您认为内蒙古环境的变化：生态环境、草场、树木被破坏，气温上升、极端天气增加

您认为内蒙古环境变化的原因：工业污染、人口排污、矿业开发

您认为内蒙古未来环境的变化趋势：有些方面会改善，人类对于环境的认识会更加理性，长期看，环境会变好.

您对于您事业的期望：第一，让公司的设备更加齐全，市场份额扩大；第二，目前还不能完全规范化生产，希望改变之前的操作方式和生产方式.

河套地区为什么不是土豆的主产区？ 因为河套气温高，无霜期长，可以种植的农产品种类多；河套是河水灌溉，不利于土豆种植；土豆种植收入相对较低.河套地区在靠近西部的沙漠地区有土豆种植，但是大规模的农场种植很少，大多是农民自产自用。内蒙古其他自然条件差的地区多种植土豆.现在内蒙古其他地区小麦亩产800-1000斤，大约1.4元一斤，农民种小麦每亩地毛收入大约1400元，投入400元，净收入1000元.木豆亩产4000斤，前几年0.55元一斤，毛收入2100元一亩，成本700元一亩，净收入1400元一亩，要超过种植小麦.现在河套地区多种植甜瓜、番茄、葵花等经济作物.

Source: Visiting Mr. Yun Ting and his company *Zhengfeng Seed Potato* 16th September 2013, Hohhot. (Photograph made by Cilia Neumann)

Source: Potato fields of *Zhengfeng Seed Potato* in Siziwang Qi, 16th September 2013. (Photograph made by Cilia Neumann)

Source: Potatoes from *Zhengfeng Seed Potato* in Siziwang Qi, 16th September 2013.
(Photograph made by Cilia Neumann)

Appendix 5

Interview with a Herdsman from Siziwang Qi

Participants: Yun Ting, Wang Yifu, Neumann Cilia and an Official from the Animal Husbandry Station (乌兰察布畜牧站) in Ulanqab.

牧民调查问卷：

个人信息：高正权

    A 年龄：50
    B 性别：男
    C 民族：汉

1. 教育程度：无
   语言：汉语、蒙古语
2. 高中您在该地区的居住时间：30年
3. 您的职业：牧民
4. 您的家庭有几口人：4
5. 您的子女数量：2
6. 他/她们的受教育程度：大学
7. 您希望他/她们将来从事的工作：*现在一个当老师，一个上大学，希望将来他们进城*
8. 您是户主吗？*是*

土地用途

9. 所有的土地总面 (亩): 3700亩
   另外还租了别人3000亩草场，向没有牲畜的牧民租，政府规定24亩地养一只羊
10. 放牧的土地总面积：3700亩
12. 您饲养什么样的家畜? *羊700只 奶牛2头*
a) 为什么您饲养这个家畜? *祖祖辈辈都这样生活*
b) 您从什么时候饲养您的家畜? *祖祖辈辈都这样生活*
c) 您是否每年都会养殖不通的家畜? *不会，主要就是养羊*
13. 你住在村庄或者你还是游牧人吗? *放牧，但是已近定居*
a.) 从您居住的地方到牧场有多远? *5–6公里*
b.) 冬季牧场和夏季牧场距离多远? *15公里*
c.) 在冬天您怎么样喂您的动物? *秋天储存草料*
d.) 您是否恐惧白灾，上一次的冻灾是什么时候？ *恐惧，上次就是去年，大约20多只羊被冻死*

环境

14. 您是否恐惧草原退化? *恐惧，会影响畜牧的质量*
15. 您认为什么是草原退化的原因? *不下雨*
16. 您为防止草原退化所做的努力? *轮牧*
17. 饲养家畜的每年家庭收入? *20万，另外还有4万元的旅游业收入，政府的补贴是1.4元一亩*
18. 您会再投资多少在您的事业上? *5-6万元*

a.) 目标? *更多的草场和牲畜*

等到他们老了，就卖掉草场，进城和子女一起生活. 现在长期雇佣一个牧羊人. 羔羊大概一年就卖. 母羊5-6年才卖. 活羊卖给商人，13元一斤，平均一只羊70到80斤. 生活的电力来源是高压电，因为距离公路比较近. 生活用水来自于地下水，平均5-6米就有水，但是有些地方地下50米都没有水，牧民只能买水，50元一次拉一车. 草场的分配，最初是按照人口来分配的，一个人分24只羊，一只羊分25亩草地，养羊只能限定在自己家的草场上. 蒙古人和汉族人生活的很好，矛盾很小，有很多汉人从事牧业，很多蒙古人从事农业. 家里有摩托车和汽车. 土地是88年分配的，30年的承包期，88年1月1日到2017年12月31日. 土地到期会延长，现在基本上都是长期的土地合同.

Source: 18th September 2013, visiting a Han-Chinese Livestock Breeder with his Mongolian wife. From left to right: Livestock Breeder, Governmental Official, Wife (Photograph made by Cilia Neumann)

Source: 18th September 2013, renting of Yurts to tourists to gain additional income (Photograph made by Cilia Neumann)

Source: 18th September 2013, storage of sheep dung for heating (Photograph made by Cilia Neumann)

Source: 18th September 2013, sheds for agricultural use, living house, restaurant yurt for tourist purpose (Photograph made by Cilia Neumann)

Appendix 6

Interview with a Taxi Driver form Dengkou, Hetao

Participants: Yang Zhen, Wang Yifu, Neumann Cilia

河套地区农民调查问卷:

个人信息:杨振

A 年龄：40
B 性别：男
C 民族：回

1. 教育程度：*中等专科学校毕业*
   语言：*汉语*
2. 您在河套地区的居住时间：*10年*
3. 您的职业：*出租车司机*
4. 您的家庭有几口人：*3*
5. 您的子女数量：*1*
6. 他/她们的受教育程度：*高中*
7. 您希望他/她们将来从事的工作: *上大学*
8. 您是户主吗? *是*

土地用途

9. 所有的土地总面 (亩): *没有地，是外来人口，从事运输业. 一年净收入2万，毛收入3万. 父母是种地的农民，居住在临河. 环境变化在变好，但是不明显. 黄河边上能捕到的鱼人少了，第一因为渔民太多，第二因为黄河水污染. 黄河边上的饭店老板大概一年收入5万.*

Source: 20th September 2013, fieldtrip to Hetao Irrigation District. Mr. Zhen Yang, Taxi Driver in Dengkou (Photograph made by Cilia Neumann)

Source: 20th September 2013, fieldtrip to Hetao Irrigation District. Wetlands near Dengkou, Yellow River (Photograph made by Cilia Neumann)

Source: 20th September 2013, fieldtrip to Hetao Irrigation District. Yellow River near Dengkou (Photograph made by Cilia Neumann)

Source: 20th September 2013, fieldtrip to Hetao Irrigation District. Yellow River, Sanshenggong Hydo-Powerstation near Dengkou (Photograph made by Cilia Neumann)

Source: 20th September 2013, fieldtrip to Hetao Irrigation District. Yellow River, Sanshenggong Hydro-power station near Dengkou, information (Photograph made by Cilia Neumann)

Source: 19th September 2013, typical vegetation and corn cultivation, gully erosion in the mountains near Baotou, Inner Mongolia (Photograph made by Cilia Neumann)